IET MANUFACTURING SERIES 24

Design Thinking for Engineering

Other volumes in this series:

Design Thinking for Engineering

A practical guide

Edited by
Iñigo Cuiñas and Manuel José Fernández Iglesias

The Institution of Engineering and Technology

Published by The Institution of Engineering and Technology, London, United Kingdom

The Institution of Engineering and Technology is registered as a Charity in England & Wales (no. 211014) and Scotland (no. SC038698).

The Institution of Engineering and Technology
Futures Place
Kings Way, Stevenage
Herts SG1 2UA, United Kingdom

www.theiet.org

British Library Cataloguing in Publication Data
A catalogue record for this product is available from the British Library

ISBN 978-1-83953-502-4 (hardback)
ISBN 978-1-83953-503-1 (PDF)

Typeset in India by MPS Limited

Cover Image : Group of co-workers working together / Gerber86 / E+ via Getty Images

Contents

About the editors

Iñigo Cuiñas is a professor in the School of Telecommunication Engineering at the University of Vigo in Spain, where he teaches courses on remote sensing and the link between engineering and society. He is the co-author of more than 200 published papers. His research focus is mainly on the propagation of radioelectric waves, and in the use of design thinking and soft skills in engineering.

Manuel José Fernández Iglesias is a professor in the School of Telecommunication Engineering at the University of Vigo in Spain, where he lectures on informatics and the social links of engineering. He has over 200 published papers. His research is focused on the application of IT to improve the quality of life of dependent and elder people. He is also interested in how blockchain can be used in education.

List of figures

Chapter 1
Introduction
Íñigo Cuiñas[1] and Manuel J. Fernández Iglesias[1]

Design thinking (DT) methodology has been developed in the later twentieth century, and it has experiment a highly rocketed impulse as a useful problem-solving strategy for many companies and institutions. Humans are in the core of the methodology, which is only understood when people is in the center of the process: the solutions proposed to the problems are made on the optics of the users instead of on the optics of the designers.

DT originated at the beginning of the 1960s of the last century in the United States, when current industrial design methods began to be questioned, mainly those oriented to the design of software systems [1]. From then on, other approaches to design were exposed, with a more scientific character, where the human component plays a more relevant role (human-centered design [2]).

In any case, it was not until the mid-1980s that the foundations of this new methodology were defined, through initiatives such as those of Professor Peter Rowe at the Harvard Graduate School of Design who published *Design Thinking* in 1987 [3], or the creator of the *Ambidextrous Thinking* course at Stanford University Rolf Faste [4]. A key milestone in the development of DT is the foundation of the company IDEO, with David Kelley as a committed promoter of the new methodology [5].

DT is the name to represent a set of approaches and techniques that were initially used by designers of physical objects or devices, but organized and structured in a way that facilitates innovation. In opposition of other problem-solving strategies, whose domain is content based, the domain of knowledge in DT is process based. This broadens the range of applications of DT to many areas of interest and to solve a large variety of problems concerned to people and/or to organizations. This methodological approach was originally applied to design, and, from there, it was adapted to the fields of education [6], engineering [7], economics [8] and management [9]. Eventually, DT began to play a relevant role in the field of education [10,11], as it can significantly improve the skills necessary to solve students' own problems, foster collaboration and broaden students' prospects. Since DT originated from design, it can even influence the design of student spaces and school systems to be adapted to the requirements of

[1]atlanTTic, Universidade de Vigo [GID DESIRE], Spain

a society in constant evolution, where innovation and people-centered interventions play a fundamental role.

DT provides tools to apply, in a systematic way, a process of observation and understanding of people (i.e., empathy) that helps to clearly identify and define the problem at hand. From there, DT fosters the generation of ideas, as many as possible, with the aim of building prototypes from the best ideas that fit the defined problem. With these prototypes, we observe how people interact with the solution designed, if other unexpected applications are found, and eventually if we were really able to solve the original problem. Along the process, if you start from the conviction that the professionals of the coming years must be able to go beyond their technical knowledge to get to understand the people who are going to use their solutions, so that these solutions are those that adapt to the people who need them and not the other way around.

DT methodology could be included among the novel agile methodologies, such as lean startup, agile and scrum, that nowadays it is most used in innovation. This methodology makes it easier for us to find solutions to a problem that it is experienced by our user or client, with totally different proposals to those that already existed. In fact, the human-oriented or user-targeted condition of DT represents a specific characteristic of the methodology. These innovative solutions generate a positive impact and, individually, we would not have been able to reach them: we need the direct contact with those people that need the solution.

As an agile methodology, it is clearly iterative, adaptive and fast. It has a very important characteristic that is the validation of each of the stages or steps followed with the user and also the decisions that are made in them. This methodology goes through five clearly defined stages and must also be applied in an adaptive or flexible way, that is, not always sequentially. In addition, these five phases: empathy, definition, ideation, prototyping and testing, must be part of an iterative process in order to improve the solutions proposed to our problem by successive approaches. With these iterative steps, we can manage project risk in a simple way.

So, the key to the success of the DT methodology is to put the user at the center of the process, and it is in this first stage of DT, called empathy, that this is clearly demonstrated. This empathic mandate, which relies on the core of human-centered philosophy, relates to a variety of methods of capturing, observing, engaging with, and immersing oneself in the livings of others. Empathy plays a double role: it puts the "human" vision in human-centered design; and it also emphasizes the dual dimension of complex problems: they could have a technical nature but they also have social dimension [6].

1.1   The DT process and stages

The DT process is generally organized in five stages [12] (cf. Figure 1.1): empathy, definition, ideation, prototyping and testing. The first two phases – empathize and define – seek to fully understand, from all points of view, the problem we intend to

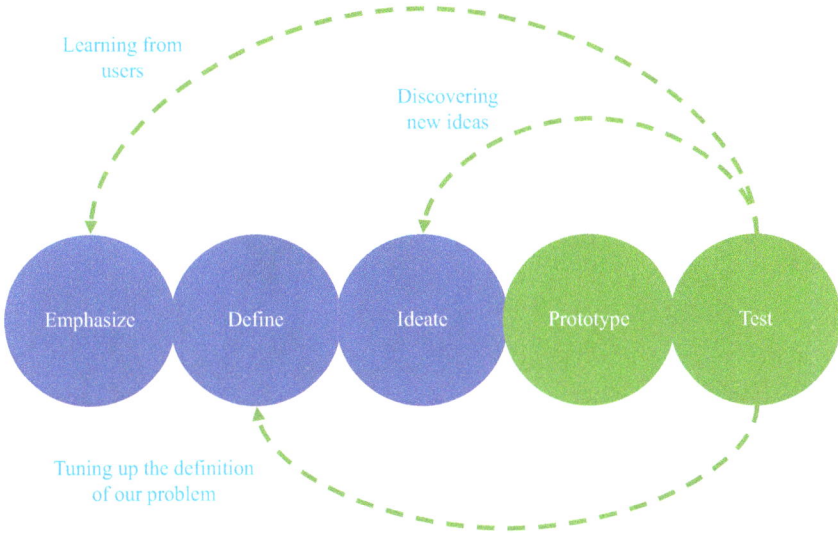

Figure 1.1   DT stages

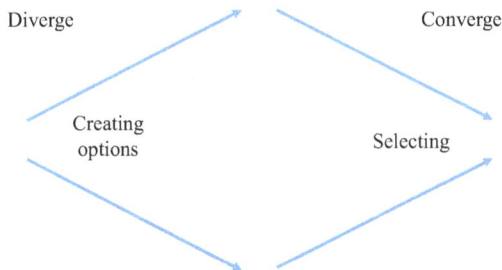

Figure 1.2   Diverge and converge process

solve and the people who experience it. The next two phases – ideate and prototype – seek to generate the most appropriate solution to solve the problem from a wide set of candidate solutions. Finally, the last phase – testing – brings us back to the actual people to whom we are providing a solution to check if the solution is indeed the best possible one for them.

The evolution of the activity along the DT process is like a succession of divergence and convergence situations that guide the design-thinkers towards proposing a solution to an initial challenge. During the divergence states, a number of options are creating, broadening the possibilities for the project development. Whereas, the convergence states are time to concretize and focus on a direction selected as a consequence of the insight from previous stages. Figure 1.2 graphically summarizes this idea, which is in the core of the DT methodology.

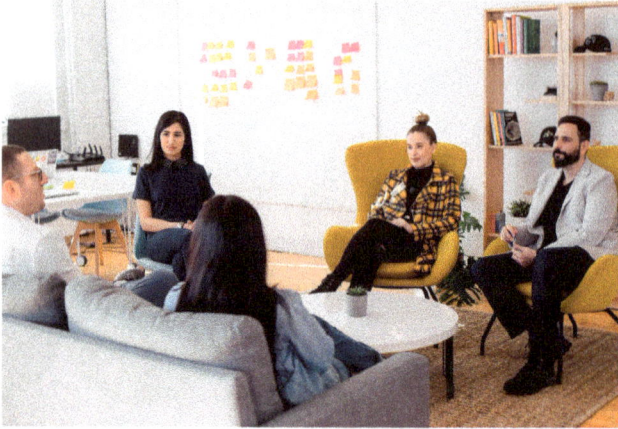

Figure 1.3   For design thinkers, users are not models or abstractions but actual persons (photo by Jason Goodman on Unsplash)

As indicated, there are five stages in the application of DT in a continuous diverge and converge solution-design process. The empathy stage aims to gather as most information as possible by observing and interviewing the possible users (cf. Figure 1.3). In this context, empathizing means interacting with people and observing their behavior and their interactions in the environment in which they are. We talk to people, and ask questions about their habits that have something to do with the problem. Collecting this information helps to empathize with people and feel the way they feel. In addition, we speak with experts related to the problem and carry out a research task using the means at our disposal (e.g., bibliographic references, information on the Internet, surveys, etc.).

This is a diverge stage, as the DT-team moves from the ignorance on users' lives and feelings to an immersive knowledge, thus opening a wide information for the next stages. Our ultimate goal in this phase is to get a large amount of information about the problem that we are going to solve, and about the people for whom we are going to find a solution to that problem.

A good strategy to empathize is to practice active listening [13] (cf. Figure 3.3). For that, try to mirror the person talking to you. A simple but effective strategy would be to paraphrase what the person to whom you are talking is saying, to demonstrate your engagement, that you really try to internalize their thoughts. As pointed out above, to better understand why someone behaves in a certain way, it is best to *put yourself in their shoes*.

Once you gathered all relevant information, it is time to unpack your findings, to process everything that you heard and saw to understand the big picture. This is also a chance to start sharing what you found with fellow designers and capturing the important parts and their relations in a visual form. For this, a common approach is to get all the information out of your head and your notes and move it onto a wall by means of sticky notes, where you can start making connections. The definition stage, the second one, aims to converge from the information gathered during the empathy to the explanation of the problem to be solved, providing a point of view (PoV), which is no more (and no less!) than a sentence that summarizes the problem indicating who needs something, what they need, and why this needing arises. It is more than describing the problem, it goes into its roots to provide the deepest reason. Then, this stage begins with a lot of information about the group of users and ends with a clearly stated definition.

The definition phase is very important, as team members try to identify and precisely determine the real needs of people, as well as the motivations that lead to feeling those needs. The strategy consists in focusing on the point of view from which the innovative idea that will solve the problem will emerge, in trying to answer questions such as *What would happen if ...?* Or *How could we ...?*

Consider what stood out to you when talking and observing people at the empathy phase. If you noticed something interesting, ask yourself (and your team) about it. Develop an understanding of the type of person for whom you are designing. Synthesize and select a limited set of needs that you think are important to fulfill; you may in fact express just one single relevant need to address. Work to express the insights that you developed through the synthesis of information. Then, articulate a point of view by combining these three elements: the actual user, their needs and the insight. This should be a guiding statement that focuses on insights and needs of an actual user, or composite character. Most of the time, insights will not just jump in your lap; rather they emerge from a process of synthesizing information to discover connections and patterns. Defining means sense-making.

The third stage is called ideation. It is again a diverging time, as different ideas for solving the problem will be its result. Using the point of view as inspiration, the DT-team provides a collection of ideas that could help in solving the problem being partial or global solutions.

While the definition stage is aimed to determine the specific meaningful challenge to address, ideation is about focusing on generating solutions to address that challenge (cf. Figure 1.4). Ideating consists of generating a large number of ideas and concepts to solve the problem, using tools such as brainstorming. The more ideas we have addressing a candidate solution, the more effectively we will carry out the remaining phases. Ideas have to be varied and imaginative, and it should be noted that, for the time being, we should not judge them. At this point, we look for ideas, without even deciding which is the most suitable or brilliant one. The process should be inspiring enough to generate a large number of proposals from which we can then select the best idea or a combination of several ideas. To complete this phase, the team must put their feet back on the ground and select one or two candidate solutions.

You ideate in order to transition from identifying problems to creating solutions for your users. You combine the understanding you have of the problem space and the people for whom you are designing with your imagination to generate solution concepts. Ideation is about pushing for the widest possible range of ideas from which you can select the best solution. The key issue is to defer judgement until the whole solution space has been thoroughly explored.

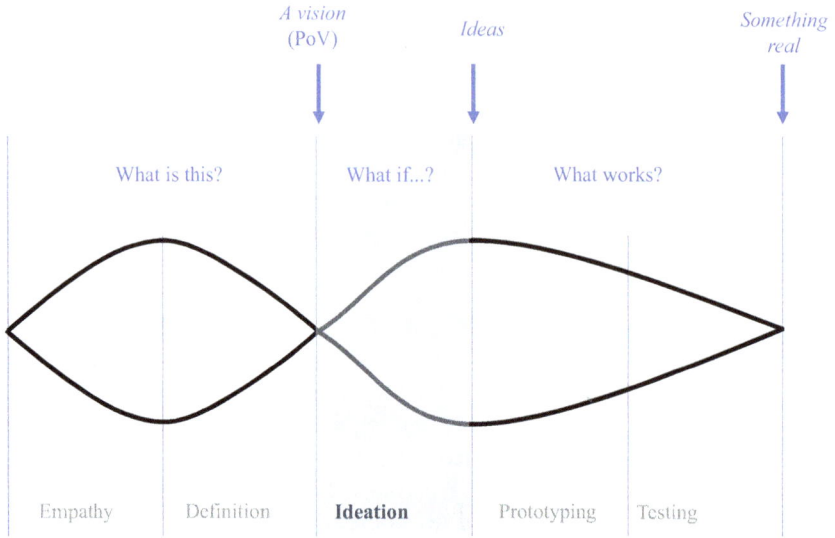

Figure 1.4   At the ideation stage, we experience another expansion process from the Point of View, which in turn was the outcome of the synthesis exercise at the definition stage

The fourth stage consists of prototyping a solution based on one or more ideas generated previously. The result of this stage is a mock-up or a simple prototype, which could be fully functional or descriptive depending on the available time and materials. Again, the result is one element, so the process has converged from a collection of ideas to one proposed solution.

Prototyping is a dynamic and very fast stage in the DT process. As pointed out above, a prototype can be just a simple mock-up made of decorated cardboard. The goal of prototyping is to quickly visualize ideas and create a playable model, which will improve communication, help to discover and avoid basic errors, and provide more inspiration. The importance of this phase lies in having, in a short time and at a reduced cost, a solution that can be tested by the people whose problems we intend to solve. It will always be more efficient to verify that we are successful with our solution in an early development phase than with a final outcome that is already in the production phase. Our strategy is to fail at reduced cost, as many times as necessary, to be successful with the final product.

A prototype is intended to answer questions that get you closer to your final solution. Prototypes are concrete representations of a system or a part of it (e.g., an object, a service, a device, etc.). They are tangible artifacts and not abstractions. In the early stages of a project, you should create low-resolution prototypes that are quick and cheap to make but can elicit useful feedback from users and colleagues (cf. Figure 1.5). In later stages, both your prototype and questions will get more and more refined.

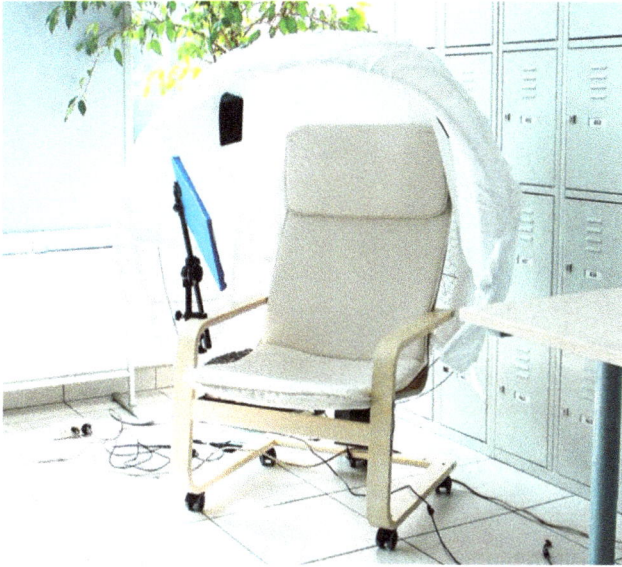

Figure 1.5 A prototype of a stress-relieving cabin for crowded areas such as airports or malls, constructed with a regular office chair and some scrap materials

The prototype is then given to the users during the last stage, the testing one. The DT team observes the interaction between users and prototype, getting feedback on its usability, utility and goodness to improve the users experience and to solve the detected problem. This stage is not clearly divergent or convergent, as depending on the gathered insight its consequence could be the end of the process (when the DT team considers that the prototype is a good response), or could move to go back to some of the previous stages.

Tests allow us to identify and understand the people's perceptions about a specific solution. In principle, nothing should be explained, since what really matters here are the observations and comments made by the people for whom we proposed a concrete solution. Testing aims to find out if the prototype meets expectations and what are the aspects or details that should be improved.

After testing the prototype, it may be necessary to revisit the previous phase to build a new prototype that brings us closer to the final solution. In fact, prototyping and testing are carried out in tandem more than as activities between which you transition. What you are trying to test and how you are going to test that are critically important aspects to address even before you create a prototype. Examining these two modes in conjunction brings up the different layers of testing a prototype. Though prototyping and testing are sometimes entirely intertwined, it is often the case that planning and executing a successful testing scenario is a considerable additional step after creating a prototype. Do not assume that you can simply put a prototype in front

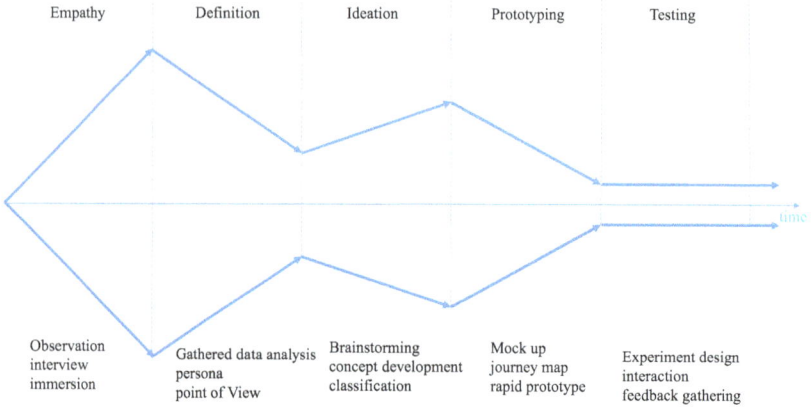

Empathy	Definition	Ideation	Prototyping	Testing
Observation interview immersion	Gathered data analysis persona point of View	Brainstorming concept development classification	Mock up journey map rapid prototype	Experiment design interaction feedback gathering

Figure 1.6   Some of the most relevant DT tools

of a user to test it; often the most informative results will be a product of careful thinking about how to test in a way that will let users provide you the most natural and honest feedback. It may also happen that we need to go back to an even earlier phase to look for new ideas or even better define the problem.

Figure 1.6 depicts the process of DT in a graphical way, including some defining activities performed during each of them, which are explained in Chapters 3–7.

Once the previous process is completed, in the case of products that will eventually be commercialized, the DT team would present its results to the final investors to proceed with their implementation from a final prototype that meets market requirements. What remains to be done is its production, commercialization and application, that is, transferring the results of the project to the business community and the general public.

Although the successive stages seem to follow a linear path, along a closed process, in fact the application of DT involves a continuous step back to previous stages in order to refine the outcomes and redefine the evolution of the process. Ultimately, DT does not have to be an inflexible and rigid methodology for problem solving. Furthermore, it should not be because otherwise it would become a more traditional methodology then losing its essentials. Stage sequencing should be used as a guide that tells us the natural evolution of events, although for each specific project, we can complete these stages in a different order, go back, perform them simultaneously or repeat them several times to broaden our view on the problem, and finally come up with the best possible solution. The information circulating throughout the stages serves to better understand our audience, the original problem, and the solutions we are planning.

A common criticism that DT suffers continuously is that DT does not really contribute anything new [14]. Analyzing the methodology, it consists on applying common sense when facing a new and complex problem. Thus, it seems to be logical to carefully listen to the audience (DT calls it "empathy"); from that knowledge, to choose with common sense the problem to face it, neither too specific nor too broad (is it "definition"?); to search for a solution ("ideation"); to make this idea a real thing or procedure (very similar to "prototyping"); and finally show it to some trusted one to receive an honest feedback on what it works and what it does not ("testing"). Obviously, we are not forbidding the application of common sense during the process, and DT is making intensive use of it like any other successful problem solving approach. However, from a methodological point of view, DT is bringing something different compared to other approaches: it provides a structured collection of methods and techniques focused on facilitating innovation. These techniques and methods already exist before DT and they can be used out of this scheme, but DT proposes a structured and coherent way for their application.

The goal of DT is not to discover an existing truth through analytical thinking, in a traditional way: this is, in fact, the role of Science! What we look for is a way of "inventing the future" by means of synthesis, after deeply analyzing the present. What is really different is that we do not expect a single "right future" but many "possible futures" among which we have to select the most appropriate, that what respond to most of our original questions.

We can estate that calm, independent thinking or introspection could be good contour conditions for generating ideas. However, DT is based on the belief that meaningful, human-centered innovations can only be developed through teamwork-based processes. This methodology is specially structured to support intense collaboration and co-creation. The creative process would flourish through the cross-pollination of multiple perspectives, ideas, and approaches, taking advantage of diversity in all its forms to break with the traditional status quo: gender, cultural, academic, functional, professional or technological.

DT is also part of the belief that the process of creating innovative products, services and experiences is inherently ambiguous and cluttered. This methodology is based on accepting non-linearity and even chaos through an open and flexible mindset, and an uninhibited attitude in favor of experimentation and play. Excessive control of the innovation process is not only useless but also counterproductive: if we want to regulate the innovative process, the solutions could be sound, interesting and even valid but perhaps not so ground-breaking or pioneering. Therefore, DT encourages a positive attitude toward uncertainty and improvisation, and confidence in instinct. We do not take ourselves too seriously, but we take what we do very seriously.

Oppositely, DT could not be a universal application methodology that will solve all possible problems, but we can say that the ultimate goal of each DT project is to design a solution that meets the following conditions:

- The focus is on people, on human-centered design. The solutions created must appeal to the needs, emotions and behaviors of the people for whom we are creating such solutions.
- Its technological feasibility must be assured. Is the solution technically possible, or does it depend on a technology that has not yet been invented or developed? Our solutions must be practical and workable without incurring unaffordable costs.

- Its social viability must be guaranteed. Will the solution work when we want to put it into practice in the real world? Is the solution environmentally friendly? Is it inclusive? Is it accessible? DT is a long-term process that should ideally continue to be supported and improved beyond the completion of the original project. Solutions must be viable and sustainable over time.
- Economic viability must be also endorsed. Will the revenue we get from any source offset the financial resources needed to implement the solution? Have we identified a sufficient flow of funds to ensure its timely maintenance? Is the project economically sustainable? In the case of a commercial product, is there an appropriate business model behind that product?

In summary, DT structures a collection of methods, tools and techniques in a way that facilitates the innovation in providing feasible solutions to problems using a human-centered focus. The process is flexible but not chaotic, it is serious but not sad, it is innovative but not mad.

1.2   DT path along its development

DT provides a modern and fresh way to face problems, but this does not mean that DT is a new methodology. Although its development has been paired with some of the most innovative companies at the later twentieth and early twenty-first centuries, its foundations come from the interwar period in the first half of the last century.

Actually, the first reference to DT, or at least the establishment of the basic ideas of the methodology, comes from 1934 when John Dewey proposed to integrate the principles of engineering and aesthetics in the making of a new generation of machines (i.e. electrical devices) based on traditional materials and old-fashioned machinery [15].

In the 1940s, the term "DT" is applied to define the state-of-the-art in mechanical engineering as a reference in the creation of new motors [6].

But the interest was not limited to mechanical engineering and, in 1957, the American Ceramic Society published a manifesto regarding the interest on improving the skills of designers learning DT proposals, considering the need of carrying design from the formation of the idea to the production line as a full "design action" with all the steps connected [16].

However, it was not till 1959 when John E. Arnold coined the term "Design Thinking." He taught classes at Stanford University and at the Massachusetts Institute of Technology and developed a course on creativity on engineering in which he began using the term that now gives the name to the methodology [17].

Another important milestone was established by L. Bruce Archer in 1965, when he defined the methodology, providing a procedural approach to training designers to add different knowledge to their original specialization: ergonomics, cybernetics, marketing and management [18]. The generalization of the usage of DT in different ambits was also supported by the book of Bryan Lawson on the way of thinking of the designers [19].

In parallel, there is a history of success that boosted DT to the important role it is now playing in the technology and engineering industries: in 1978, a new design firm was established under the name DKD (David Kelley Design). This consultancy company could be identified as another one in their ecosystem, but they applied DT as a dogma from their beginnings... and Steve Jobs asked DKD for designing a mouse for the new Apple computer at that time, named "Lisa." They applied DT to design the mouse, which was included slightly modified with the first Macintosh computer. The rest of the history is part of our modern lives.

Also in the early 1980s, Bill Moggridge led the design of the first notebook-style computer for GRiD systems. The design team incorporates a new practice of observation of users interacting with the computers software that they called interaction design. The know–how of the companies these visionaries, Kelley and Moggridge, was merged in 1991 into a new company that incorporated also Mike Nuttal firm, creating IDEO, which is now the paradigm of DT-based development. IDEO success, and their innovative designs, increased the interest for DT from many engineering and technology sectors in later twentieth and early twenty-first centuries.

Besides this industrial awareness, academics provided additional support and research on the DT proposal. The same year when IDEO was born, a workshop meeting was held at the Faculty of Industrial Design Engineering, Delft University of Technology, The Netherlands. This was the first symposium fully dedicated to research in DT, and its contributions are available at a book published in 1992 [20].

There were a succession of interesting research papers linking the new proposals of DT with those original ideas of John Dewey, as the well-known work from Richard Buchanan in 1992 which frames the challenge for DT: the contribution of design to the modern world. Buchanan connects DT, wicked problems and Dewey's theory: "What Dewey defines as technology is not what is commonly understood in today's philosophy of technology. Instead of meaning knowledge of how to make and use artefacts or the artefacts themselves, technology for Dewey is an art of experimental thinking" [21].

In 2005, the Stanford University, which was leading the DT movement, erected its Hasso Plattner Institute of Design (better known as the d.school). This institution, the d.school, created and offered courses on DT. Besides, its full-time staff is researching, improving and promoting DT at different education levels [6].

Different people at Stanford University and IDEO have influence in popularizing DT as a valuable method to be learned at many different fields outside professional design. DT is useful for innovating in business, universities and organizations of any type. The idea that anyone, form different fields of knowledge, can become a design thinker results to be very democratizing [22].

Within this inspiring environment, technology and engineering could identify DT as a beneficious methodology in their missions. And this is the starting point for writing the present book: explaining the process within the field of technology, providing tips for implementing courses at university level in faculties of engineering, of different fields of engineering, and also sharing successful experiences on the use of DT methodology with pre- and post-graduate students in a collection of disciplines. Prestigious researchers, as those cited but not in a comprehensive history, provided the society with several tools and theoretical support regarding DT; now, we aim to apply at least part of this knowledge in our engineering fields.

1.3   Inspiring DT examples

To complete this introductory material about DT, you can find below some relevant real-world application examples belonging to a broad range of application fields.

1.3.1   The embrace story

Every year, around 15 million premature and low birth weight babies are born in the world. In developing countries, the mortality of babies is very high due to the lack of incubators in hospitals, among other issues. Today's incubators are extremely expensive, and the incubators available in these hospitals, generally older and discarded by First World hospitals, are often difficult to maintain, operate and repair.

A group of students from the Stanford School of Design that were participating in a Design for Extreme Affordability course studying technologies suitable for people living on less than one dollar a day faced the challenge of identifying the reasons for high infant mortality in developing countries. They realized that one of the most pressing problems was regulating a premature baby's body temperature. The team began their research in Kathmandu, the capital of Nepal, and after spending several days observing the neonatal unit of a Kathmandu hospital, they went to investigate how premature babies were cared for in rural areas. They were aware of the need to observe babies in their natural environment, and they encountered two situations that caught their attention. First, the majority of premature Nepalese babies were born in rural areas, and second, most of these babies were delivered to hospitals because their mothers were unable to care for them.

To save as many lives as possible, any viable solution would have to work in a rural setting far from a hospital, and it would have to be easy to use by rural mothers. It would also have to run without electricity and be transportable, intuitive, hygienic, culturally appropriate, and, perhaps most importantly, relatively inexpensive. The result was a kind of textile warming bag [23] composed of three elements: an infant-sized sleeping bag, a pouch of phase change material (PCM, a substance that releases or absorbs enough energy to generate useful heat or cooling at a phase transition) and a heater. The pouch, when warmed in the heater and placed into a compartment of the sleeping bag, maintains a temperature of 37 °C for a period of up to 8 h.

The team took the prototype to India, where they tried to understand the cultural nuances that might lead mothers to accept or reject the device. They discovered issues that they would never find out at their homes in Silicon Valley. For example, when one of the team members was in a small town in Maharashtra showing the prototype to a group of mothers, designed to complement skin-to-skin care, he told them that they should heat the bag that they designed to 37 °C to help regulate the temperature of the baby. This received a surprising and disturbing response. One of the mothers in the village explained that in her community they believed that Western medicine was very powerful and very strong for them. For this reason, if the medical staff prescribed any medication for their children, they would give them half the dose. Similarly, mothers will heat the bag to thirty degrees Celsius or so, to keep their little ones safe. To avoid this, they modified the design of the warmer so that when it reached the correct temperature, a visual indicator will be activated. This way, it would not be necessary to deal with any numerical temperature value. This is a good example about how a simple modification of the initial prototype, improved the final solution in a way that could mean the difference between life and death.

After completing the design course at Stanford, the original team continued to evolve. New members joined, they improved their business plan, expanded their funding sources, and in 2008 founded the non-profit organization Embrace. In 2012, they launched a for-profit company, Embrace Innovations, to manufacture the warmer. The company has developed two versions of the warmer, marketed as Embrace Nest and Embrace Care. Embrace, the non-for-profit organization, continued to focus on improving health care for babies in less developed countries. In 2015, Embrace joined Thrive Networks, an international NGO working to improve the health and well-being of marginalized communities in Asia and Africa through evidence-based technology and programs.

In 2016, the infant warmer manufacturing was transferred to Phoenix Medical Systems in India. Clients come from private clinics, public-health facilities, and NGOs. Embrace warmers were used to care for the more than 200,000 preterm and low birth weight infants in 19 countries: Afghanistan, China, Ethiopia, Ghana, Guatemala, Haiti, India, Kenya, Malawi, Mali, Mexico, Mozambique, Nigeria, Rwanda, Somalia, Sudan, Tanzania, Uganda and Zambia.

1.3.2 Domestic toilets

The IDEO.org team designed a comprehensive sanitation system to meet the needs of low-income families in Ghana [24]. The Clean Team service is a custom designed freestanding rental toilet, as well as a waste disposal system. In addition, the design work was extended to provide a complete ecosystem of services, including branding, uniforms, a payment model, a business plan, and key messages. Unilever and WSUP piloted the project with about 100 families in Kumasi City, Ghana, before launching it in 2012.

One of the key aspects of the success of the project was a consequence of the DT's empathy stage. The work team interviewed experts in sanitation, shared work sessions with Ghanaian operators of public toilets and interviewed a significant number of common people. This allowed them to get a clear understanding of what the ideal toilet would look like in that particular context and what would be the most appropriate way to collect waste.

Later, several prototypes were built, some of them as modifications of existing portable toilets, applying what was learned from end users and other stakeholders in the previous stages. These prototypes were put to the test by real people. With this, they were able to make a decision on the best toilet design and also obtained invaluable information on how to offer an attractive service and promote it. With the testing of the different prototypes, they discovered a key aspect that went unnoticed until then: the scarcity of water was going to be a determining aspect when proposing a definitive toilet model and when deploying the service that would be made available to the Ghanaian people.

1.3.3   DT on rails

In 2016, the French public railway operator SNCF adopted DT to enhance passenger experience at its stations across France. Following the path taken by other big companies in the past, SNFC's motivation relied on the fact that high-quality design together with usability was instrumental to develop an emotional connection with customers and users. The DT's solution-oriented approach to achieve innovation, by putting the customer or user as the central element in all development stages, was key for SNCF to position itself to compete in a rapidly evolving transport market.

Since then, several DT projects, at several scales and degrees of deployment, were developed through Gares & Connexions, an SNCF subsidiary responsible for 3000 French stations serving 10 million users every day [25]:

- Employees in charge of satisfaction surveys were equipped with a simple yellow badge displaying message *your opinion interests us*. With this simple measure, survey participation increased, as respondents perceived questioners as friendly and concerned.
- In France, specially in the case of high-speed trains, departing platforms are announced just minutes before the time of departure. This increased stress among travelers and negatively affected the whole traveling experience. The suspension of giant balloons displaying the hall number above the station concourse enabled passengers to rapidly identify departure points.
- The *En Gare* mobile application provides passengers with suggestions about what to do at the station, according to their location and the time available before departure.
- Cardboard cut-out characters placed in isolated areas at stations are used to comfort passengers traveling alone at night. This was proven to offer a friendly presence and an understanding that help is not far away.
- Baryl, a robotic dustbin, was deployed at Paris Gare de Lyon to circulate the station and move towards passengers looking for a dustbin. Upon receiving any waste, Baryl issues a polite *merci*.

1.3.4   Keep the change

This is a project developed for the Bank of America to promote a culture of savings [26]. IDEO.org came up with a new service for which the customers of that bank could sign up: a savings account that over-rounds debit card purchases. Then, the difference between the actual expense and the rounded expense is automatically transferred to another savings account. In addition, the bank remunerates the money transferred at the second savings account in a special way.

As is customary in this type of projects, design thinkers were integrated into the Bank of America offices in several cities in the United States. From that position, they were able to interview a large number of men and women to learn first-hand about their experiences in relation to household finances and banking habits.

As a consequence of these observations, they realized that in many cases mothers took care of household finances. At a time (early 2000s) when many families still used checkbooks to track their expenses, it was common practice to round up expense entries to facilitate account keeping, but it also provided a small margin for savings. This observation was the key to the new business model.

With this project, *small change accounts* based on automatic savings transfers started to proliferate all around the globe. Apart to its contribution to personal savings, the idea was applied to other scenarios such as charities, college funds or microfunding.

1.3.5 A really smart toothbrush

When Braun and Oral-B asked the Industrial Facility design studio to design the ultimate smart toothbrush, a people-centered design strategy led designers to propose two actually useful functionalities: to use any available USB port to recharge the brush's batteries and that users were notified when they have to change worn-out brushing heads [27].

This project was launched in the mid-2010s, at a time when Internet connectivity was already ubiquitous in developed countries, and it seemed that any new gadget had to be connected to the network to be successful.

In many cases, the new connectivity options, together with the availability of inexpensive miniature sensors, had as a consequence the proliferation of new services that, although technologically advanced, did not have a real demand or acceptance by real users. Everything had to be smart, spawning a mindless bubble of connected devices that lasts to this day.

An approach to design based on DT was instrumental to detect what was really appreciated in a toothbrush, namely the two functionalities identified at the beginning of this section.

1.4 Organization and contents of the book

After introducing the DT process in its five stages, describing some historical facts on the development of the methodology and introducing a collection of inspiring experiences with DT, the interest of this book in engineering education must be clear. This section will introduce the contents of the following chapters, and also describe the structure of the book.

The first part of the book is devoted to the general view of the methodology and the description of its different stages, including a collection of tools and activities to perform with students.

The second part shows different proposals and experiences at diverse engineering fields, explained by the lecturers in charge of them.

Then, the third part focuses on how to incorporate DT in engineering curricula, including a possible course guide and a collection of good practices when teaching this methodology.

1.4.1   DT and its stages

The first part of the book is a guide for lecturers and students to learn about how to implement DT for project-based learning courses, including the description of the different stages and a collection of tools that would help in developing each phase. It consists of six chapters, from Chapters 2 to 7. Chapter 2 presents a proposal for having a fast and immersive experience of developing a simple project on DT basis. In fact, 2 hours would be enough to go over all five stages and advance in the proposal of a solution. This trial could be motivating for diving deeper on the methodology, once the reader has checked the enormous possibilities it provides. The design of a DT laboratory is also included in this chapter, as the author assumes that it is important what we do when applying DT, but the environment where we do DT could also help in reaching good and creative solutions.

The focus of Chapter 3 is the empathy stage. This is probably one of the trademarks of DT process, as from the beginning the users, the people are in the center of the development. This idea is sometimes complex to be understood by engineering students, who are more used to apply solutions to previously defined problems and not to define problems related to people and then look for solutions. In fact, it is more than understand: the design thinker has to deeply feel that the user is more important than the problem to be solved. Having this in mind, the chapter defines the basic elements of empathy, highlighting their interest and characteristics.

A reflection on the innate condition of empathy or the possibility of becoming empathetic by training and learning is also included in this third chapter, as a support to those that do not think that they are close to the people. Finally, a collection of tools to improve empathy and to perform empathic interactions with other individuals are also described in this chapter. Eventually, the reader will have the tools to perform an empathic immersion among users to gather as much information as possible on their feelings, lives and needs.

Definition stage is the target of Chapter 4, that is, the explanation on how to identify the essential challenge and how to identify the actual problem which solution will be addressed (i.e., to define the problem). The information gathered during the empathy stage should be converted into insights that provide the keys to define the problem to be solved. Finding a good "point of view" is instrumental at this point, that is, a motivating and memorable definition of the problem that moves the DT team to look for solutions. An explanation of these basic tool is the main content of this chapter.

Besides, other supporting tools for the definition stage are included in the fourth chapter, namely the concepts of stakeholder and persona, the empathy maps, the efforts and results charts or the affinity maps. These tools could be useful in the task of organizing the gathered information and converting it into insights related to the users.

Once the reader has acquired knowledge on the users and their problems regarding the actuation problem scenario, the next step is how to generate ideas to solve these problems. Chapter 5 is devoted of this third DT stage: ideation. This chapter includes considerations on how to ideate and how to promote the generation of ideas among individuals working as a team. Important elements as thinking out of the box, lateral and combined thinking, question assumptions, exploring the extremes, changing the focus or inspiring questions are among the contents of this chapter.

Ideation relates to creativity, but generating an idea for effectively solving a problem requires some methodological background. Thus, different tools are included along this fifth chapter in order to support this creative process. They are introduced and some activities are proposed to learn how to apply them in engineering education.

Chapter 6 deals with prototyping. This is a very motivating DT stage for those learning the methodology, as they have the possibility of constructing their proposed solution, which thus moves from the world of the ideas to the physical world. The objective is to embody the outcome of the DT process in a physical object, in order to have something tangible to be shown to the users to check the validity of the result. The chapter contains information regarding the characteristics of a prototype, its role and the aspects to be taken into account: form, fidelity, interactivity and evolution.

Different rapid prototyping techniques are introduced and explained along the sixth chapter, providing a variety of procedures to build the prototypes (sketches, mockups, videos, wizard of Oz) to allow the reader to select those that better fit the proposal. Besides, a selection of team challenges to develop skills among students to improve their prototyping abilities are included at the end of the chapter.

Once we have a prototype, the next step is to observe what users do with this object. This is the aim of Chapter 7, the fifth stage of DT: testing. The chapter justifies the needing of testing as the moment when design thinkers really understand if the problem under study is solved by their idea or not; i.e., if the users manage the prototype in the way it was thought or they interact in a different and unexpected action. The chapter explains how to organize the testing phase, using the observation to get user feedback to improve the solution to the problem.

Along this seventh chapter, a collection of tools that can be useful for the testing phase are introduced and developed: presentations, infographics, feedback capture grid and storytelling are the proposals to share with DT learners and improve their testing skills.

At the end of this first part of the book, both lecturers and students would have acquired a deep understanding of the DT methodology, as well as a collection of tools and exercises to create routines among the future design thinkers in the development of a project. The experiences shared at the second part will complement and complete this knowledge. Finally, the third part will help lecturers to organize these tools and exercises in courses of different format.

1.4.2 *Experiences of DT application in engineering*

The second part consists of three chapters in which authors share their experience of applying DT together with students to different problems in electrical and telecommunications, mechanical and biomedical engineering. Different approaches lead the reader to discover what kind of solutions can be reached by design thinker teams, even when they are novice learners.

Chapter 8 shows the experience of a course on engineering and society. The proposal, after providing a short practical workshop on DT, was to move the students to different areas of a Spanish regional airport with the mission of improving the experience of the travelers in such facility. The chapter describes the methodological approach, how the engineering students discovered the actual needs of the users and how the airport managers involved within the proposal.

The results of the experience within the airport are also introduced along Chapter 8, organizing this part into the different projects developed over the experience. The authors also included the way they assessed the works and some lessons learned during the development of the proposal.

The automotive market is the focus of the experience described at Chapter 9, involving Polish students from Mechanical Engineering. The lecturers describe how they taught their students, the process of empathic research and problem definition, and all the stages of DT methodology. They give special attention to the point of view in the definition stage, and on the process of creative ideation on the proposed future innovation for automatic cars. The developed ideas, the created prototypes and the insights discovered at testing stage are also part of the contents of this ninth chapter.

Chapter 10 deals with the application of DT in bioengineering. The specific characteristics of this discipline make it really interesting for applying DT, as bioengineering products interact directly with users, in fact with patients. So that, the process of empathy results to be especially critical for the final solution.

The chapter analyzes the different DT stages when experimenting with students and real-world users, paying special attention to the empathy step but considering all of them. Besides explaining the author's DT experience in her field, she also provides some reflections on what are the main aspects when using DT in bioengineering.

This second part of the book results to be descriptive regarding the possible use of DT in engineering education, and it could be motivating for lecturers that wanted to incorporate DT ideas in their lessons, as all these proposals have been developed in subjects of not more than 5 months as a complement of other traditional contents. Thus, they are eye-opening for those doubting on moving to a more active way of teaching.

1.4.3   *Incorporating DT in engineering curricula*

The third part of the book is focused on supporting lecturers in organizing a subject based, or partially based, on DT methodology. Chapter 11 contains a course guide, Chapter 12 is a collection of tips and lessons learnt regarding DT in the classroom, and Chapter 13 is a reflection on the effect of DT on students' creativity.

On the one hand, Chapter 11 describes all the contents of a typical course guide in the European Higher Education Area, but easily adapted to other education systems. It includes objectives, skills or competencies to be acquired and learning outcomes but also the course syllabus, the planning, the methodologies and the assessment procedure. It is written in a way that facilitates the adaptation to DT course as a part of a wider subject or as a subject as is, considering the length and requirements of the project as the factors to adapt the total credit allocation. This chapter would help lecturers to create the academic definition of a project/problem-based learning subject based on DT methodology.

On the other hand, Chapter 12 focuses on how to deliver the DT course in a practical way: tips for each of the DT stages, tracking the students' teams, different formats for the course, interdisciplinary teams and lessons learnt. The main contents are the reflections of teams of experienced DT lecturers after teaching some years courses on DT at different levels, having the aim of helping DT newcomers when, once defined their courses, they go into the lab or the classroom to lead the course. This hands-on experience should be valuable to decide how to perform the different stages and how to deal with previously reported situations.

Finally, Chapter 13 contains the reflections of a very experienced professor on how DT helps in boosting creativity among students. After explaining his views and the historical views on creativity, the idea racing system appears as an interesting proposal to improve the capacity of generating ideas and, what is as important, to keep them ready to be use when needed.

The value of DT as a vehicle of creativity is one of the key ideas of this 13th chapter. The creativity related to innovation, but also related to self-esteem and personal happiness, arises as one of the future pillars of our education and our society.

This last part of the book will, then, help lecturers to design and to deliver a course based on DT ideas but also to understand the deep effect of DT practices on students' lives, not limited to their intellectual knowledge or their social skills.

Part I

Design thinking and its stages

Chapter 2
A first taste of design thinking
Manuel J. Fernández Iglesias[1]

The world is full of challenges that need to be addressed. In this context, innovation plays a fundamental role in today's society, but experience shows that it is not enough to build a better world. It is necessary that innovative solutions to our problems are also sustainable and accepted by the people, in addition to being technologically feasible and economically viable. Design thinking (DT) is a methodology called to contribute to fostering a favorable attitude towards responsible innovation among students and professionals related in one way or another to innovation, and also to inspire companies and institutions to implement methodologies and instruments with the goal of fostering teamwork, innovation and people-centered creativity.

DT is a methodology used by leading companies such as Apple [28], Virgin [29] or Toyota [30] that demonstrated its suitability to promote innovation in organizations in an effective and successful way. Steve Jobs' decision to adopt DT when he returned to Apple in 1997 is widely credited for setting the company on its path to today's privileged market acceptance and to rank at the top of the most valuable brands because of its cutting-edge design and a sense of high-quality, but without detracting from usability.

DT is aimed at promoting and developing people-centered innovation, offering a series of instruments to identify real challenges and problems in order to eventually solve them through innovative proposals. In this process, empathy, ideation techniques, rapid prototyping and tests in real environments play a fundamental role.

DT uses the sensitivity of design professionals and their problem-solving style to meet people's needs in a way that is technologically feasible, socially acceptable, and commercially viable. This solution-oriented approach attempts to achieve innovation by putting the consumer at the center of all development stages. As a side effect, DT solutions strive to develop an emotional connection with the people using them.

The rest of this chapter is devoted to provide a first taste of DT by means of a 2-hour activity designed to experience all the DT stages that does not require any previous knowledge about the methodology. Additionally, we outline some ideas about the design and provision of a suitable space for design thinkers to carry out their projects, that is, the DT Laboratory.

[1]atlanTTic, Universidade de Vigo [GID DESIRE], Spain

2.1   Design thinking: a 2-hour journey

This is an activity that allows you to experience the complete cycle of the DT methodology in a short period of time, without even requiring prior knowledge or familiarity about it. This activity is based on the DT tutorials from the Hasso Plattner Institute of Design at Stanford University.* As discussed here, it is an adaptation of a proposal developed within the framework of the DiamonDT project, funded by the Executive Agency for Education, Culture and Audiovisual (EACEA) of the European Union within the Erasmus+ Program.

   This activity is designed to be done in pairs. It only requires a pen or pencil and something to take notes, and it can be done in any space that allows participants to interact comfortably and without interference.

   It is advisable to have the possibility to play music in the background at specific times. As a general rule, music will play during working periods, in pairs or individual work, and it will be paused when mentors or tutors provide explanations or clarify some aspect of the activity, or when participants present their work. Anyway, keep in mind that music must function as background music, that is, it must allow the participants to work comfortably.

   The specific topic of the activity must be presented just before starting it, never in advance, and participants should be instructed not to look at the information of the next stages of the project before completing previous stages.

   It is important to be strict with the activity times. The working time must be measured with a stopwatch (e.g., the one in a smartphone is perfect), and the end of each stage must be announced with a strong signal, such as a bell, a gong, a whistle, and a siren.

   The roles played by the members of each pair alternate throughout the exercise, so that each of them will be able to experience the roles of project developer and project target.

   The activity uses as supporting materials a series of paper templates. It is also convenient to have a computer presentation ready where information on each step of the DT process is provided and the activity templates are displayed while participants are working with them. This presentation will serve to familiarize the participants with each stage of the methodology while carrying out the activities of that stage.

   The experience is organized according to the DT stages: empathy, definition, ideation, prototyping and testing, and begins by proposing a specific work topic. These topics should serve to provide some focus without restricting too much the imagination or creativity of participants. There will be time, throughout the experience, to choose a specific problem to solve (through the point of view) and provide a specific solution to it (through a prototype).

   Some examples of topics to carry out in an extreme DT experience could be:

- Get rid of one of your vices.
- The best way to get from A to B (choose A and B to be inspiring to your audience).

*http://dschool.stanford.edu

- My tribulations with clothing.
- How would you improve the experience in the lavatories of your work or study facility?
- How would the perfect Sunday look like?

An interesting variant when proposing work topics would be for the participants to assume the role of an imaginary, famous or historical person who is known enough to enrich the phases of empathy and problem definition. This means that, at least in principle, participants do not have to reveal any aspect of their real personality, although it limits the authenticity of the experience. In this context, examples of topics to work on would be:

- James Bond needs a food processor.
- Dressing Luke Skywalker.
- Redesigning the bat-cave.
- The perfect time management tool for Wonder Woman.

Cards can be printed out representing these project proposals and have each pair choose one at random before starting the experience. It is not necessary to have ready as many different projects as participating pairs. It would be OK if different pairs are assigned the same project.

The total duration of this activity ranges between 1 and 2 h, depending on the time spent on phase shifts and the time spent on final presentations in the testing phase, which in turn depends on the number of participants. These Extreme DT sessions can serve as an introductory session to a broader DT course or workshop, as a demonstration session to introduce the methodology, as a special session in a course on any subject where project-based learning is used, etc.

2.1.1   Empathy

In this phase, participants will interview their partners to get as much information as possible about them in relation to the assigned project. Each member of the pair interviews the other in turns, and the first interviewer within each pair should be announced at the beginning of the exercise, for example by asking the participants who will conduct the interview first to raise their hands. This makes it easy to keep track of time and role switching. For the rest of the experience, each pair will have a perfectly identified A & B members.

The objective of this stage is to try to empathize with the other member of the pair, knowing details of his life, needs, aspirations and desires. It should be noted that this is essential to build the best solution to solve a specific person's problem. The questions should be open and should allow exploring different paths throughout the interview. For example, questions beginning with Why …? or a What for …? are preferable to those that require a yes or no answer, such as *Do you have a car?*

Each participant has 4 min to find out as much as possible about their partner. After that, roles are swapped. Figure 2.1 illustrates a possible template for this phase.

After the two members of each pair were interviewed, we pause for a short minute to review the responses obtained.

Time: 4 min

1. Interview your partner an get as many information as possible.

Notes from the first interview.

Switch roles and repeat.

Figure 2.1   Extreme DT: template for the first interview

Time: 3 min

2. Interview your partner again and complete the information gathered.

Notes from the second interview.

Switch roles and repeat.

Figure 2.2   Extreme DT: template for the second interview

After the first round of interviews, a second three-minute round per participant is conducted in order to deepen our knowledge about our partner. It should be pointed out to participants that this is a good opportunity to deepen their knowledge with deeper questions or to explore interesting aspects discovered after the first round of questions. A template proposal is depicted in Figure 2.2.

Once both interview rounds are completed, each participant will spend some time working alone capturing, analyzing, and recapping their findings. The objective now is to start the definition stage, so it is convenient to introduce the concept of

Time: 4 min

3. Capture, analyze, summarize.

Goals & wishes: What does your partner try to achieve? Use verbs!	Insights: What can you see in your partner that not even them can see? Infer from what they told you.

Figure 2.3   Extreme DT: template for the closing of the empathy phase

point of view (PoV) and its importance (cf. Section 4.2). Each participant will record their observations and discoveries made during the two rounds of interviews on the appropriate template. They should also write down the goals and desires of their partner. Note that the description of the objectives has to be done using verbs: my partner says that ..., my partner feels better when ..., my partner needs ..., my partner loves ..., etc. Figure 2.3 depicts a template proposal for this exercise.

In short, this stage consists of the following exercises:

1. First round of interviews (4+ 4 min).
2. Pause (1 min).
3. Second round of interviews (3+ 3 min).
4. Analysis and summarizing (4 min).

And the total time to complete it would be around 20 min.

2.1.2   Definition

In this phase, we continue with the definition of the point of view initiated in the previous stage. Each participant, individually and for 3 min, identifies and selects a problem or need that, in their opinion, is the most important to their partner. The PoV must clearly indicate the person for whom it is defined. With this, we emphasize the fact that we are dealing with actual people. In the case of the variant based on role playing and for the duration of the activity, participants become the character whose role they assumed. Now, they are real people that we have to be taken very seriously. The most common approach to identify the protagonist of the PoV is to write their name, that is, the person referred to in the PoV that each participant is defining. Figure 2.4 presents a possible template for this phase.

Time: 3 min

4. Formulate your partner' point of view (PoV).

(Name of your partner)

needs

because/but/surprisingly....
(choose one)

Figure 2.4   Extreme DT: point of view's definition template

Like all PoV, it must be concise and must identify the essence of the problem. Therefore, it must explicitly identify the need addressed and the justification or insight to address it, for example:

- Marisa needs to avoid the Christmas shopping chaos because she has four children and a lot of things to do before Christmas.
- Antonio has just become a father for the first time, and he needs to connect with other new parents because he often finds himself lost and isolated and needs to feel that he has everything under control.

For this exercise, we do not explain any existing technique to address PoV definition. We trust the participants' instincts and intelligence.

2.1.3   Ideation

The goal of this stage is to outline five innovative ways to meet the needs of the other pair member based on the defined PoV. To do this, we provide each participant with a template with five blank spaces (cf. Figure 2.5).

Before starting, participants are briefed about the power of images. They are told that drawing pictures is often much more valuable than using textual descriptions. In addition, participants should be encouraged to be creative and radical, to *get out of the box* and be daring when drawing. The ideas outlined may be simple, but they should offer a functional solution to the problem or need defined by the PoV. To make this process more efficient, the PoV will be copied in a specific box provided in the template used to collect ideas. This allows participants to constantly observe the point of view, and thus focus on creative work without wasting time looking back to find the content of the previous template.

Time: 5 min

5. Draft at least 5 innovative ways to satisfy the needs of your partner.

PoV: _____

Figure 2.5   Extreme DT: template to collect five inspiring ideas

Time: 5 min

6. Share your solutions with your partner and get feedback.

Your notes.

Switch roles and repeat.

Figure 2.6   Extreme DT: ideation feedback template

The duration of this exercise is 5 min, but a small competition can be organized at this point by inviting participants to generate a collection of solutions as quickly as possible, with a time limit of 5 min.

As in the previous stage, we do not propose any specific technique for ideation.

Once the ideation exercise is finished, participants work again in pairs to present the ideas generated to the other team member and obtain their comments. The information gathered from peers is recorded in a specific template (cf. Figure 2.6).

Time: 5 min

7. Propose your final solution using the new information.

Describe your final solution. add all necessary details.

Figure 2.7   Extreme DT: final idea collection template

The importance of constructive criticism and its impact on the quality of the final solution should be emphasized. Participants should be encouraged not to be afraid to criticize and to be criticized. They should also include in their notes the deficiencies and discrepancies between the solutions discussed and their own vision about the needs they must satisfy.

The participant presenting their solutions should write down the feedback they receive, which will allow them to choose the best solution from their partner's point of view, improve it, or find a new, better one. The feedback obtained in this step can also help participants to update and improve the defined PoV. The time allocated for this task is 5 min per participant.

After the feedback exercise is completed, participants go back to work individually to create their final solution. They should try to include as much additional technical details and descriptions as possible in a specific template provided for this purpose (cf. Figure 2.7).

The solution sketch and its description will be the guide to build an actual prototype. Nevertheless, in a setting where time constraints and working conditions do not allow to carry out a full prototyping session, participants may draw a sketch together with the appropriate comments, as a simple 2D rapid prototype. The time allocated to this task is 5 min.

To sum up, this stage consists of:

1. Outline of five solutions (5 min).
2. Feedback (5+ 5 min).
3. Creation of the final solution (5 min).

For a total duration of around 20 min.

Time: 10 min

8. Build your solution.

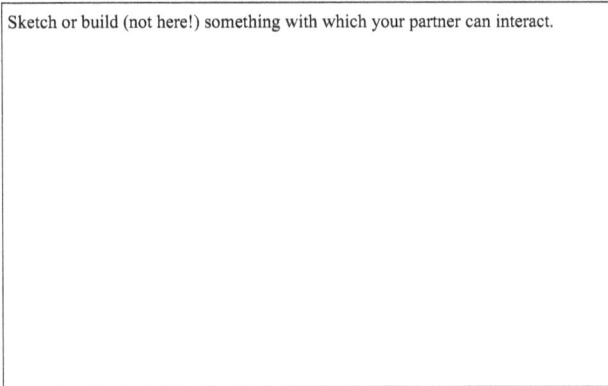

Sketch or build (not here!) something with which your partner can interact.

Figure 2.8 Extreme DT: 2D prototyping template

2.1.4 Prototyping

The prototyping stage should be carried out using materials and tools available in the place where the extreme DT experience takes place, such as office supplies in the case of a classroom. No specific template is required. However, in the case that a sketch is deemed as an adequate prototyping technique, such template may be provided for the sake of completeness (cf. Figure 2.8).

Participants shall be encouraged again to be creative. It should be insisted that the prototype, its functionality and its details, must reflect the functionality that solves an actual problem and satisfies an actual need of the other member of the team. Participants have 10 min to complete their prototype. Note that rapid prototyping is a key element with real value in the field of DT.

Depending on the time available, it may be interesting to organize a prototype exhibition or fair.

2.1.5 Testing

In this last stage, participants present their prototypes to the rest of the audience and collect their reactions using the template provided (cf. Figure 2.9). During prototype presentation, participants should be encouraged to ask questions and provide constructive criticism. Thus, the presenter has the opportunity to obtain additional valuable information that could be used to further improve the prototype.

Each participant will have 8 min to make their presentation. In any case, depending on the time available, exhibitions can be limited to a single member of each pair, or to a limited number of participants chosen at random, chosen by the rest of participant, by the people who act as tutors or mentors, etc.

Time: 8 min

9. Present your solution and get feedback

Pros/likes	Cons/drawbacks

Doubts	New ideas

Figure 2.9   Extreme DT: feedback collection template for the testing phase

2.2   The design thinking laboratory

A proper work environment can dramatically improve the experience of those involved in problem solving using DT. This requires a suitable physical space and the right equipment and materials. The editors have conducted workshops and courses for teachers, university students, and the general public for almost 10 years. The proposal below comes from that experience.

First, the physical characteristics of the laboratory itself are identified. Then, we present the basic equipment required to conveniently develop the different activities throughout a DT project. Some additional equipment is also suggested that may improve the participants' experience. Finally, we list the essential materials in every reasonably equipped DT laboratory.

2.2.1   The physical space

A workspace where people can feel comfortable will make them feel happy, which in turn promotes productivity. As in other teamwork methodologies, it is convenient to have a sufficiently large space where design thinkers can work in groups and move around comfortably. DT projects often require all participants to get up and engage in a certain activity in front of a board or panel on the wall. In addition, mentors or facilitators also need space to carry out their tasks. They must be able to move freely around the lab to keep work teams engaged and motivated.

Chairs should be light so that they can be moved easily, and they should also ensure good posture. Instead of large tables, smaller ones that can be arranged in different ways are preferred. We must not forget that the amount of furniture is limited by the fact that there must be enough space to move freely for any table or chair configuration.

Some furniture is also needed to conveniently store and organize equipment and materials. A toolbox or plastic box with a lid to transport materials during field work or within the laboratory is also advisable.

Finally, if possible it would be a good idea to have a designated space for snacks and drinks (water, coffee, tea, etc.).

With regard to wall decoration, having the walls painted in colors is not mandatory, but it helps to create a special atmosphere, so that participants feel involved in a different process than what happens in a traditional classroom or office.

2.2.2 Basic equipment

Once you have identified your perfect venue, it is time to equip it so you can address there the most enriching DT challenges. For this, our experience tells us that there are certain elements that you should definitely consider.

2.2.2.1 The board

The lab needs at least one whiteboard, panel, or wall device or area of the wall where design thinkers can write to complete many of the typical project activities, such as participant boards, affinity maps, and empathy maps. Enough free space is also required for a workgroup of four to move comfortably in front of the whiteboard. Suitable writing material in different colors are also needed, as well as appropriate erasers or cleaning elements to wipe the board clean. We will prefer boards able to held sticky notes and those that can be erased better than devices that cannot be erased, such as easels with large sheets of paper.

Note that large windows or glass walls or doors can be used as great writing boards, as well as for sticking notes. They are usually easy to erase or clean and provide a very convenient surface for sticky notes.

2.2.2.2 The projector

A projector facilitates the exchange of documents and information among the members of the workgroup. It makes it very easy to jointly review a document, visit a website, or share web tools. On the other hand, a projector ensures that all team members are using the same information. In addition to the projector, the appropriate cabling and connector and a suitable projection surface are also required. In some cases, one of the whiteboards can be used as a projection screen, which greatly simplifies the wall space requirements of the laboratory.

In labs with network access, the latest projector generation supports screen sharing and direct projection of some file formats from a USB stick, network disk drive or from the cloud. These features drastically reduce the cabling needs and the need for a computer to host the documents to be projected.

An alternative to projectors are flat screens. They can be placed on the wall or on one of the tables in the laboratory, and they usually have the same connection capabilities as state-of-the-art projectors.

2.2.2.3 The camera

A camera is needed to take pictures of evolving prototypes or the outcomes of wall activities. This makes it easy to reuse sticky notes and general project documentation. Note that team members working in a project can always use the built-in camera in their smartphones.

2.2.2.4   Additional equipment

Besides the elements discussed above, it may be convenient to also have the following available:

A **regular color printer:** is useful for transferring information from the digital world to the real world. For example, images taken during interviews or graphic information downloaded from the Internet can be printed out and attached to participant boards.

A **3D printer:** is a great tool for rapid prototyping. Plastic-thread 3D printers are reasonably priced, and the quality and cost of printed objects are suitable in most cases for a DT prototyping phase.

A **coffee machine/water boiler:** coffee breaks are great for maintaining a relaxed and healthy atmosphere within the group. In addition, an occasional caffeine kick can dramatically improve productivity.

A **portable audio player/speaker:** music makes people less stressed. When people feel better and less burdened, they are more creative. In addition, music facilitates mind wandering in the positive sense, that is, to gain distance from the activity at hand, leaving our thoughts to flow freely, which in turn increases creativity. There are many playlists with music suitable for DT activities on online services such as Spotify or YouTube.

2.2.3   Laboratory materials

The list below collects all the basic elements required to carry out the key activities throughout the different phases of a DT project. Although all items on the list are considered a must, they are ordered according to their relevance to the DT methodology.

Sticky notes: Sticky notes are used extensively. They are essential for transferring information from our minds to the physical world. They are needed to gather information from interviews to create participant boards, build affinity maps or empathy maps, or during brainstorming, among many other tasks. We should have a good assortment of them in a variety of colors and sizes. The most convenient size is the standard 76 mm × 76 mm (3 × 3 in.).

Felt tip pens: The perfect complement to sticky notes. Due to the thickness of their writing and their ink density, text written with them can be viewed from many different angles and distances. They are also perfect for avoiding verbiage when writing on sticky notes. We should have a reasonable supply of them, at least in black, red, blue, and green colors.

Markers: The ones suitable for writing on white boards or plastic or glass panels. It is good to have both fine-tip and a coarse-tip ones. Coarse-tip markers are very convenient for highlighting text and graphics, while fine-tip markers are good for drawing more detailed storyboards and text.

Large roll of wrapping paper: It is very convenient for building participant boards, affinity maps or empathy maps, etc. Sticky notes stick very well to wrapping paper, and marker pen writing is very legible. It is also extremely useful for prototyping.

Pencils and pens: It is most convenient to have available some HB pencils with their corresponding erasers, and some pens in various colors. They are very appropriate for taking notes and for quick sketches on a sheet of paper or a notebook.

Scissors: Consider having a box of four or five blunt-tipped scissors circulating around the lab.

Sticky tape: It is needed for sticking information on walls, building prototypes, or even sticking notes on surfaces where they do not stick on their own, such as some whiteboards. Consider having clear tape, duct tape, masking tape, and double-sided tape.

Glue stick: Glue sticks are very clean and are primarily designed for gluing paper and cardboard, and they are not as strong as liquid adhesive. Presently, permanent, washable, acid-free, non-toxic, solvent-free, and tinted varieties are broadly available. A box with several sticks circulating around the lab would be very convenient.

Gomets: They are very suitable for classifying sticky notes, for labeling things or people during activities, or for voting. Consider having a good assortment in various colors (e.g., blue, yellow, red and green) and shapes (e.g., round, triangle, square).

Paper and cardboard: They are very suitable for prototyping and also for making posters, cards, badges, etc. Consider including a variety of papers, plastic wrap, and cardboard sheets of different sizes, thicknesses, colors, and textures in your lab kit.

Stapler: In addition to joining paper or cardboard, staplers can also be used during prototyping to join other materials.

String: Thread, twine, or fine rope can be very suitable for prototyping, for joining elements in a board or for marking spaces on the wall or floor.

Building blocks: They help participants to innovate and express their ideas visually. They are very convenient for rapid prototyping. For example, a basic model can be built using blocks, and participants can improve or modify it as creative thinking flows. Furthermore, moving these models and modifying them adds a fourth dimension, time. As models change, the solution space evolves. It would be nice to have a variety of blocks in different sizes, shapes and colors.

Modeling paste or clay: This is a great complement to the building blocks for rapid prototyping. Simple models of real-world objects can easily be made using modeling paste and other simple materials such as paper, cardboard, or pieces of string.

USB sticks: These items are less and less relevant due to the ubiquity of network access. However, a couple of them will still be useful because they are very convenient for transferring information from participants' computers to a projector, or for sharing information if network access is not available. A capacity of at least 4 GB should be sufficient in most situations.

In addition to the materials in the previous list, our experience indicates that there are many other materials that were useful on some occasions (cf. Section 6.5): rubber bands, blocks of expanded polystyrene (EPS foam), hard wax pens, lip clips, white glue, blades, lighters, balloons, ethyl vinyl acetate sheets (EVA rubber), colored pencils, wooden shoulders, drinking straws, toothpicks, aluminum foil, contact glue, universal glue, clothespins, corks, cardboard tubes, plastic cups, paraffin candles, etc.

Chapter 3
Empathy

Íñigo Cuiñas[1] and Itziar Goicoechea[2]

Identifying and describing a challenge is the starting point to apply design thinking. That challenge should be defined in a broad way in order to open our minds and it is commonly a problem related to people. Many times, the challenge is proposed or defined by the institution, the company, or the people that asked us to find a solution. A challenge is not exactly a problem, but from the challenge, we have to interact with users to define the problem to be solved, and then to look for possible solutions.

In this chapter, we will discuss the first stage of the implementation process, empathy, perhaps the key piece of the DT methodology. Figure 3.1 depicts the main issues of this stage within the full DT process. In this first phase, it is intended to know and understand the challenges to be addressed, taking into account the environment in which the user finds himself or herself. You will see the different elements to consider in this first stage, and we will make a small reflection on a non-negligible matter: Am I empathetic or can I train to be empathetic? Finally, we will describe a series of tools to promote empathy among our students and environment, so that they can experiment and train them.

3.1 Definition and elements of empathy

Empathy is the ability to put ourselves in other people's shoes, to feel how they feel, in order to interpret their needs, their concerns, their tastes and their interests. As the American educational psychologist Carl Rogers would say, "to be empathetic is to see the world with the eyes of the other, and not to see our world reflected in his eyes" [31].

The objective of the empathy phase, also called research or collection of information in an empathetic way, is simply to identify what is really important about the target persons or users, what their daily problems are, how they interact with the environment, etc. The activity of looking at people around us gives us insights into what they think and feel, and provides us an idea of what they need and what their most

[1]atlanTTic, Universidade de Vigo [GID DESIRE], Spain
[2]Industrial Engineering School, Project Management Area, Universidade de Vigo [GID DESIRE], Spain

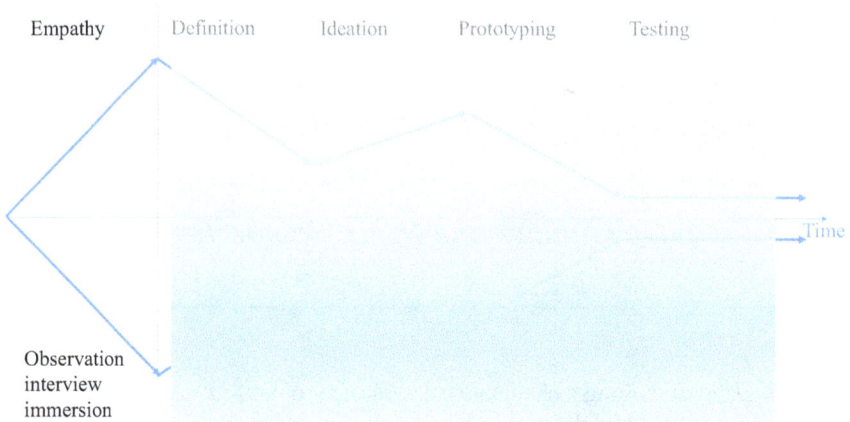

Empathy   Definition   Ideation   Prototyping   Testing

Time

Observation
interview
immersion

Figure 3.1   Empathy within the DT process

important problems are. Experience tells that, eventually, the best solutions are provided by human behavioral understandings. Those insights represent a very important concept in the DT methodology that will arise later, in the explanation of the definition phase. They are defined as a need, a desire or a belief of a well-differentiated person.

The point is, it is not so easy to carry out empathy. Taking the words of Alfred Adler, "to look with the eyes of another, to listen with the ears of another, and to feel with the heart of another" must not be an easy task [32]. Besides, it is a time-consuming activity, even this time should be identified more as an investment than as a waste, considering the benefits the project development will obtain. In order to do this, first of all we must put ourselves in the place of the other persons, knowing what their problems are in the day to day, observing their environment and what they need to solve, and second feel them as ours, that is, be one more of them.

We have to understand the reasons why people do what they do. And we must be clear that empathy is a divergent phase and that implies being able to open the mind. To achieve this, it is necessary that we adopt an open minded attitude, without making judgments of what we hear, see or feel, and this is the only way to be able to empathize. Not making judgments is an empathetic action, and we know how difficult it is, to be present, to listen and not to give an opinion. It is an additional challenge in this first phase. Remember that this phase is a phase of divergent thinking, this means that you have to open the range of possibilities and always keep in mind the open minded mentality. It is not a phase to express an opinion or say what needs to be done. If in this phase some solutions arise to you, you should write them down for other phases, but looking for solutions is not the objective of this phase, as this approach could make us misinterpret or bias our observations towards those initial ideas. We have to adopt a candid, open mind.

Another aspect to take into account as a key in empathy, according to Daniel Goleman, is to capture both verbal and non-verbal messages from the other person [33]. The rational mind is transmitted through words, and the emotional mind through body language. It is very important to know how to interpret the signals that the other person emits unconsciously and usually non-verbally. Recognizing these emotions is the first step to understanding them and identifying with them.

At this point, we can talk about two types of empathy: affective and cognitive empathy. In affective empathy, I feel what you feel. It has to do with the ability to feel another person's emotions, feelings, and sensations. In cognitive empathy, I understand what is happening to you. That is, we know how you feel and we understand it. We must develop both of them in this first phase of DT.

There are people who are empathetic by nature, for them it is something natural and even innate. These types of people have a good stretch done in this first phase of DT. But also to say, that these people watch that an excess of empathy or hyper-empathy, is not appropriate either, since it leads us to assume the emotions of the other as ours and suffer them, and can prevent them from seeing clearly the person observed. In this case, it means being a mirror and in turn a sponge. As in life, excesses are not good and the ideal is to seek balance. For those who do not present a natural empathy, this skill can be worked, like all social skills, and with time and constant training it can be perfectly developed, and even lived.

To start with this phase of empathy, the first and most common thing is to identify the users related the challenge. That is, to identify all the possible people involved in the problem that is being treated. We called them the stakeholders of the project, product, service or process of our challenge. If we want to know what a user thinks, we must first identify the different types of users, who may have different perceptions of view. This identification will be useful to be able to observe them in their environments and then to choose the most representative ones for the realization of the interviews. Sometimes, we are only able to identify stakeholders later, when analyzing the empathy insights during the definition stage. In this case, we would have to deal with all people found at the challenge environment.

Then, once the users are identified, it is time to perform the empathy stage. The three most important elements to acquire empathy and put them into practice in this first phase of DT are: observation, empathetic interview and immersion, in growing order as graphically described at Figure 3.2. These are the topics of the following subsections in this chapter.

3.1.1 Observation

Observation is a user-centered technique. It is also called covert observation since we immerse ourselves in the user's ecosystem going totally unnoticed. It is one of the oldest techniques used in research, and yet it is very useful. It aims to get to know a person or a group of people and understand the situation they are in. It is not about seeing around, but looking at the persons carefully and their environment, and for this we use a method, which will allow us to learn to observe.

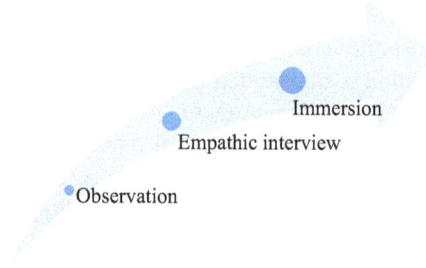

Immersion

Empathic interview

Observation

Figure 3.2   The three main elements of empathy

To do this, it is good to ask yourself three basic questions: "what...?,", "how...?,", and why...?. Moreover, within the question "why...?" it is highly recommended to use the 5 why technique, which is one of the most widely used problem-solving techniques introduced by Toyota Motor Corporation in the 1950s [34]. What do we get with these three questions?

What? It allows us to get closer to what an observed person does. With this, we realize what happens in the scene. We must not forget to write down all the details, trying to be objective. We do not want to create a story around what we observe, so we should not assume anything, just see what is happening. This means that the response of this question is pure description of the situation.

How? It helps us understanding the scene. We ask ourselves how the observed person acts, we see if that activity requires any effort, if he seems happy, hurried, hurt, if the activity impacts him or her positively or negatively. In this phase, it is good to write with descriptive phrases and with many adjectives.

Why? With that, we interpret the activity of the person observed. Do we know why he does what he does? Why does he do it in that way? It is the time of riddles and presumptions, based on the motivation or emotions that the observed person seems to have. It is about adding meaning to the observed situation.

Sometimes, if it is possible and you have authorization, you could use the camera study technique, which allows you to see the day to day of the user through their own eyes. In this case, it would no longer be a covert observation. With this proposal, the person under observation carries a camera which records his daily activity.

In this phase, it could also be interesting to obtain information about the challenge or topic that we are dealing with through experts or looking for different types of information of interest. This type of research is what we call documentary research.

3.1.2   The empathetic interview

After the phase of observation of people, in which we see their behavior and attitudes, it is time to establish contact and interact with them. It is very important that at this moment, we are aware of the value of the time those people are sharing with us in the interview. Since time is a very precious and limited element, we must make the most of it, since in many cases, we will no longer be able to be with those people and interact with them. That is why it is very important to prepare for interviews. Below we name a series of recommendations or tips.

- It is necessary to draft the questions, but in addition to know how to ask questions, you have to know how to listen. How many times do we string one question after another and not listen to the answers? Silences are also very important, so that the people interviewed take their time. Listening to the silences, and to the answers, is a first step to concatenate adequate questions: we are doing an interview and not a survey. Doing an interview goes further than asking successive questions. This means that we should change the order, omit, or add questions with the aim of keeping the attention of the interviewee and to maintain the tension of the interview.
- Remember to use powerful questions like "What?," "How?" and "Why?" as if they were a mantra. Also, the question "Why?" is very important as we will see later, and pay attention to "technique of the five whys?" as a way to find the deepest root of people feelings and needs.
- Never use targeted questions, such as: Do you think this...? Be aware that interviews are not to validate your hypotheses, but serve to understand what people think.
- We must at all moment make the interview natural, and nothing forced. Let it seem spontaneous, even though we have previously worked on it and we have clear the objectives. Remember that our goal with the interview is to get to feel what the interviewee feels in his environment.
- Try to get people to tell you their story, apply the technique of active listening: nodding your head and looking into their eyes makes the interviewee feel heard, and important.
- Do not forget to capture the results and if you have permission to video recording the interview, much better, since you catch all the non-verbal language that in an interview sometimes you are not able to gather.
- Make a summary report of each interview. From here will come the insights of the second stage of DT, the definition.

Interviews can be approached in different ways, but the usual strategy at the beginning of the interview is to start with more superficial questions thus creating a more relaxed atmosphere. Throughout the course of the interview, and once we reach that degree of confidence, we can move on to deeper and more detailed questions. In general, we can talk about three types of interviews: structured, semi-structured and open.

Structured interviews: are very rigid research tools. Standardized interviews can be considered and have a quantitative approach. Previously, a planning of all the questions and the order in which they are going to be carried out is defined. The questions are always closed, so the people interviewed do not make comments or appreciations about the questions, and are limited to giving a concrete answer.

They are useful for getting information for a large amount of people, although you can lose part of the freshness of the process.

Semi-structured interviews: are based on a script with the issues that we want to deal with the person interviewed. But, although we follow a basic structure in the interview, it gives some leeway when probing the interviewees. The questions will be of an open type and the interlocutor can show opinions, nuance answers, even deviate from the script when unplanned topics that are of interest appear. It is necessary to have an ability to naturally introduce important themes, linking topics and answers, which needs a bit more experience than when applying structured interviews.

Open or unstructured interviews: also called in-depth interviews, they do not start from a script nor from a list of questions. They are more like a conversation, but having an objective along the chat. The person conducting the interview has a great responsibility to identify at each moment what you have to ask and how to do them, and get the best answers. They are usually interviews conducted in several sessions. The success of this interview is the ability of the interviewer to achieve a climate of trust or harmony with the interviewee. This type of interview already requires a certain practice. We can consider them as an art.

Analyzing these three types of interviews, we can say that, in the case of our empathetic interviews, we will not rely on very structured interviews with totally closed questions, but we are more inclined to semi-structured interviews or even open ones, when the interviewer already reaches a certain degree of practice in this task.

What should the questions look like? They must pursue to create that climate of trust, seek proximity to the interviewee, and thus the interviewee can feel free to respond openly. Remember that it is necessary to plan the interview, creating a list of possible questions, and organizing their rhythms and their depths, starting with superficial questions until we reach the deepest ones.

All this selection of questions is obtained after previous studies, from the results of the prior observation process, or by performing a question brainstorming technique among the participants in the DT process. Once we have that list of questions, we will organize them by topics and redefine the statements to create a natural communication, so that the interviewees do not feel questioned. You have to prepare how to link some questions with others.

Now, we are in the step of analyzing if we have enough questions, we have questions of the type what, how and why or a new question that we will talk about later, which is for what? We have enough direct questions about how the person we interviewed feels for breaking the ice. Never ask multiple questions at once. The key is to perform them one by one, and wait for each response. If it is not clear, ask a counter-question. Once this previous work is done, it is time to start the interview. We have previously selected the people we are going to interview, after analyzing what kind of people are involved in the project and choosing the most representative. Before starting the interview, we need the agreement of the other part, hence the importance of that first contact with the interviewee, which is crucial in creating a receptive environment.

We already have the agreement of the people we are going to interview and we already have a possible list of questions organized by topics and superficial questions to create that climate of trust at the beginning of the interview, so we only have to choose suitable place and time.

Check-list (Interview preparation phase)

1. Analyze the information obtained in the observation phase
2. Select interviewees (representative of stakeholders)
3. Seek the agreement of the interviewees (it is the first contact with them, and we look for them to be receptive)
4. Create list of questions by topics. Check if you have:
   (a) Superficial questions for the beginning (objective to break the ice)
   (b) Questions like What?
   (c) Questions like How?
   (d) Questions like Why and For what?
   (e) Open questions
5. Redefine the list of questions (to transform into natural conversation, spontaneous, connection between questions)
6. Select interview location
7. Select the time of the interview
8. Select interview material
9. Select the way you dress

It is important to select a suitable and quiet place where both interviewer and interviewee are comfortable and relaxed. And as for the selection of the moment, it is important to emphasize that an interview requires time and full dedication if we want it to be successful, so we must look for a time when it is not conducive to interruptions due to the work of the interviewee. When an interview is initiated, the interlocutors should be focused on it and have nothing urgent to do next as it can make or lose mindfulness in it. In addition, the interviewee must know the agreed time that the interview will take, it is even good to finish a few minutes before.

For what? vs. Why?

Let us talk about these two types of questions. They are two ways to ask when we want to know more about the motivations of doing something. And we usually ask them in these two ways: "For what?" or "Why?" When preparing an interview is it good to make a small reflection about what you feel when you are asked a question that begins with "Why?" and what do you feel if that same question starts with "For what?"

Why? questions are usually related to the more theoretical motivation that explains your actions and your feelings. It is something more descriptive and more passive.

On the other hand, For what? questions generate action, it is something more active and motivating, as responses indicate what you intend to do with your actions or feelings.

About the material we need for doing an interview, at least a pencil or pen, a field notebook and the script of the interview. In the case that we have authorization to record the interview, a voice recorder, although a video camera is always more advisable, which will allow us to analyse posteriori, in addition to the conversation, the non-verbal or body language of the interviewee.

Another aspect that we must not forget is to select the way of dressing because we must adapt our clothes to each interviewee and each specific situation: i.e. dressing in a suit and tie to talk to a student can create distancing. The goal is to create a climate among equals, allowing a success interview face to face.

Once we have everything ready to start the interview, we analyze the different phases through which the interview must pass. To do this, we can observe Figure 3.3 in which we represent on the horizontal axis the duration of the interview, and on the vertical axis, the degree of attention and emotional intensity.

Presentation: The first thing we have to do is introduce ourselves. Saying who we are is the first step to building trust and stopping being a stranger. It is very important these first minutes, which allow the interviewee make a composition about you. Hence, it is essential to do it in a relaxed way and with a smile, and always in a natural manner. The attitude we must adopt is assertive, respectful and open. Our goal is to promote communication and initiate that relaxed climate.

Presentation of the project: Then, although in the preparation phase of the interview, we had a first contact with the interviewee, seeking his or her agreement to participate, and you have already explained the reasons for the conversation, it is very important to reinforce the objective of the project. It is necessary, therefore, to explain to the interviewee what you do, what you are working on, why we need

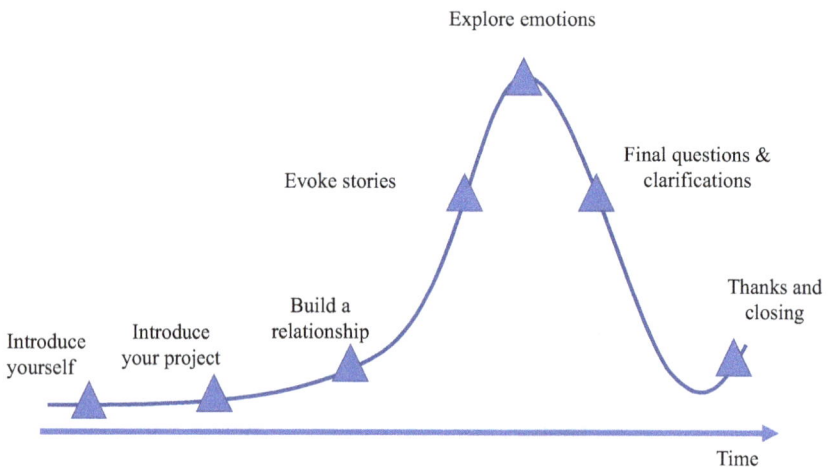

Figure 3.3   Empathy. Emotional intensity at each of the phases of the interview.

their answers and, what they will be used for. At this time, the language we use must be clear, without much technicality by which you can be overwhelmed, and adapt it to the profile of the interviewee.

Building a relationship: This phase is essential. From it begins the highlight of the interview in terms of the degree of confidence and arouse interest in the interview. But how do we create that climate of trust? For this, we use questions and counter-questions. Mention again the importance of going to the interview with a way of dressing according to the interviewee and the specific situation, to ensure that there is no distance between the interlocutors.

Evoke stories: What is sought is to identify situations in the life of the person interviewed that have to do with the objective of the research.

Explore emotions: At this moment, in addition to telling us their experiences, it is necessary to get their feelings out of him, when making them.

Perform counter-questions: During the interview, when listening to the answers, we will have doubts. In this case, it is important to ask counter-questions to make sure that we understand everything that the interviewee answers us.

Thank you and close the interview: Finally, it is very important to convey to the interviewee the importance of their answers, thanking them for their collaboration. We will close the interview by saying goodbye to him or her, but perhaps it will not be a definitive goodbye, if we have a second session planned. At this time, the conformity of the date, time and place of the meeting is sought. Sometimes it may be required by the interviewee to read the transcribed interview to complete nuances to certain questions. That is something that we have to agree on and it can be very interesting.

As we can see in Figure 3.3, at the beginning of the interview the degree of emotional intensity of the interview is very low. The first questions are of the closed type, to collect information of the demographic type, such as gender, age or work activity. But as the interview progresses, we change the type of questions to questions of the open type, as the ambient is more relaxed and it is at this time when the degree of emotional intensity rises until it reaches its maximum in the exploration of emotions. Then there is a small phase of decay with the counter-questions and questioning, to go back up slightly with the final thanks.

Interview's goal

Our goal in the interview is to keep the person interviewed emotionally active, seeking their maximum involvement and complicity in the interview.

During the interview, you should instill comfort, asking open questions that allows the interviewee to extend. Questioning him the reasons for his story encourages him more to tell his story, to delve into it and by continuing to show your interest in it, the person ends up showing how he felt.

Your body language is also very important: you have to show that you are totally dedicated to the interviewee. It is required be mindfulness and practice active listening.

The world has stopped, and in this moment, nothing distracts you. You are focusing on taking notes and listening every word and every gesture or expression, nodding your head, expressing that you are fully attentive. You have to convince yourself that the key person in the interview is him, and not you, the one who asks, and we must demonstrate this from minute one.

Active listening

Active listening is a very powerful tool and above all, highly recommended in interviews. It consists on listening to the interested party with the five senses, totally focused on what they are telling you, with full attention. It is not about passively listening to the interviewee; it is feeling him in your body.

A good practice would be to observe a person in active listening and see and analyze what they do. What will we observe?

The person who actively listens conveys interest in the interviewee, and we can see that through verbal and non-verbal messages that he performs, such as maintaining eye contact, nodding with his head and smile, showing agreement with affirmative expressions (yes, aha, ok, etc.). That feedback we give to the interviewee (verbal and non-verbal) will make them feel more at ease, feel important, and will make the person interviewed communicate more easily, and the communications will be more open and honest.

Non-verbal language is very important in this stage of active listening:

- A sincere and natural smile shows gratitude by making us participate in their speech.
- Eye contact, which involves listening with the eyes.
- Body posture, tilted slightly forward while sitting listening, shows interest towards the person.

However, all these expressions must be natural, never forced, since they can have an opposite effect than expected.

Verbal language is the perfect complement to active listening. The most effective approach would be:

- Remembering things during the conversation such as the name of the interviewee, some confidence, details of previous conversations, etc.
- Asking for relevant aspects or asking for clarifications that show that we are interested in what they say.
- Repeating or paraphrasing what we just heard to show that we understand it.
- To summarize the idea you have just transmitted. Summarizing involves identifying the main points of the message received and repeating them in a clear and logical way, and allows the interviewee to correct any necessary detail.

Remember to avoid any sign of distraction on your part, such as looking at the phone, your watch, doodling on a notebook, showing restless attitude or showing to be in a hurry.

As an activity to practice these skills, it is proposed that each person be interviewed by two other members of the team, so that while one asks the questions, the other takes notes about both verbal and non-verbal answers. Body language or non-verbal language transmits much more than we imagine. There are gestures and emotions that we usually go unnoticed while we maintain a conversation, and they have a lot of information. The idea would be: "listen to what is not said and observe what is not done."

Another very interesting option is to record an interview between two people, and then see it again in a group and write down each person, the verbal and non-verbal language. Put it together, analyze it and propose points for improvement.

Once the interview is over, start working on it as soon as possible to have recent impressions and sensations, and not forget about them. If the interview is recorded we will listen to it again, and we will review and put in order the notes taken during the interview. This fieldwork also allows us, in the case of a second interview, to tackle again aspects that were not clear or that require further elaboration.

Interpersonal interviews with proximity

Below we present, as an example, possible questions with proximity to listen to the perception of people who use the services of a shopping centre located on a university campus. The objective of these questions is to gain their trust, and in this way get information about how the users of that shopping centre live.

- Are you a student? Do you work at the University? Do you work in a nearby company?
- How often do you visit this shopping centre?
- What do you think about the state of the facilities?
- What businesses are missing in this shopping centre?
- What time do you have each day to eat, have a coffee?
- What do you think about the possibility of bringing food from home?
- Do you prefer quality or good price for having lunch or breakfast?
- Which payment method do you prefer?
- Do you define yourself as a classic or innovative person to eat? Why?
- What do you think of self-service restaurants?
- What was the last time you ate well is these restaurants?
- And the last time did you eat badly?
- Do you like to have lunch more by daily menu or by order like in a restaurant? Why?
- What is your opinion of the variety of menus offered by the locals? Do you miss something?
- Do you think people with food intolerances are taken into account? And what about ethical or religious food conditions?
- How would you rate these premises based on quality/price? What price would you pay for eating?
- What is your favourite place? Why?

- What reasons do you find in favour to eat or have a coffee in this shopping centre and not in another next one?
- What reasons do you find against eating or having a coffee in this shopping centre and not in another next one?
- Finally, in case that you were the owner of one of the locals, what changes would you make? For what?

3.1.3   The cognitive immersion

It is one of the most powerful techniques used along the empathy phase, in which you now become the user. With all the information previously gathered during the observation and the interviews, we are already prepared to live the experience of putting ourselves in the shoes of the people we have just interviewed.

Its potential is due to the fact that it places us in situations lived by the user, and thus we will understand him better. Imagine that we are involved in a challenge to improve the experience of blind people, so I will have to experience how a blind person lives. I will have to blindfold my eyes and experience what he lives in his day to day.

There are two ways to apply cognitive immersion. The first is to experience what an individual lives spontaneously. From the notes taken during the observation stage on the day to day of the user, we will replicate them and discover the eases and difficulties that the user lives, generating a great impact on us, because we live it and we feel it. The second could be called a planned cognitive immersion, through a journey. This is done after the interview with the user, and identify those parts of the day that produce the most important moments. Thus, we create a calendar of experiences to live those most important.

Thanks to this technique we will be able to detect the problem and the insights that surround the user in next phase.

Immersion in the shopping centre

Following the previous example, we can practice immersion by having a coffee in the different cafeterias of the shopping centre. The most popular and the least one. We can tour the shopping centre with a circuit planned by the most relevant points, use the different entrances and exits, use the bathrooms, etc.

Therefore, immersion is not simply collecting the information gained in observation and interviews. Immersion requires a deep analysis, in order to get into the skin of our target audience. For example, in the immersion in the shopping centre that we have just described, we need to study in depth the movements of all the people through the shopping centre, to define that planned circuit, which is really representative of all the people who use the shopping centre, and thus be able to make that representative route and experience what they feel.

3.2 Am I empathetic or can I become empathetic?

Reminding the definition of empathy, it is the feeling of identification with someone or the ability to identify with someone and share their feelings. It is one of the pillars of emotional intelligence and is related to understanding, support and active listening.

With empathy, we understand a person's feelings and emotions even when he is having a hard time. We should not confuse it with other emotions such as compassion, since in the latter case the person, in addition to putting himself in the place of the other, also tries to put an end to his suffering. That is, to be compassionate you need empathy, but having compassion also implies ending the suffering of the other, while in empathy, this does not happen.

Empathy is one of the most important social skills in our daily lives, and like any social skill, there are people who bring them as standard or at least have more predisposition to have them at birth, and others who need to work them continuously throughout life.

People who are empathetic identify with the following characteristics. Do you identify with them? That means that you are empathetic.

- They are sensitive and have the ability to feel what others feel.
- They like to listen actively.
- They are able to interpret nonverbal language.
- They are respectful and tolerant of issues or responses they disagree with.
- They do not believe in extremes. When they listen, they do not place themselves at the extremes but they try to look for the middle ground.
- They presuppose the goodness of people until proven otherwise.
- They try to express themselves without causing negative impact on the other.
- They consider each person different, and they act according to their circumstances.

But you should not be concerned if you do not consider yourself an empathetic person. There is a solution to this. As we know, these kinds of social skills do not appear in the academic curricula and are increasingly demanded in all areas, so it is highly recommended to work on them at all time, and whenever we have the opportunity. If you do not have empathy, this will not imply that you cannot participate in a creative DT process, but with constant work you can develop this skill and become empathetic. So it does not have to be a limitation to work with the DT methodology. Here are some tricks to be empathetic:

- Develop active listening.
- Temporarily pause your own judgments and criticisms. Do not make judgments.
- Follow healthy guidelines. Be aware at all times of the other person's verbal and nonverbal expressions. Answer properly and show interest in what he is telling you.
- Remember that empathy takes practice. Look for moments and situations where you can practice empathy on a daily basis: friends, co-workers, family and even people you do not know.

The following section provides some tools to develop and train empathy, which could help in this training.

3.3   Empathy tools

In order to develop empathy, we have to train it. Thus, in this section, we present some activities that will help you to develop empathetic behaviours. We will divide them into three sections: observation, interaction and interview. To carry out these activities, it is very appropriate and advisable to develop them as a team, or at least as a couple.

3.3.1   Observation

For this first activity, we propose to draw someone from the team. It will be done in a limited time. That is why we will get in pairs. First, this activity will allow us to break the ice and create a climate of trust with the rest of the team members and secondly to start working on observation. It gives very good results when starting any course and allows to create a relaxed climate.

Activity 1. Depict your partner
We divide the group into pairs, and ask them to draw each other for a time of 2 or 3 minutes. If it is possible, divide couples in such a way that they do not know each other, or do not know each other well.

Once the pairs are made, they are asked to make a quick sketch of the person they are paired with, using a pencil and paper we give them. Tell them not to worry about making a work of art, because this is not the goal. Try to reflect what they see.

After the appointed time, we ask participants to sign their drawing and give it to the person they drew. Next, we ask participants to comment on their drawing and why they did so and, we ask the people drawn to give their opinion on how they look drawn, whether they look represented or not.

This first activity creates a very positive environment in the team, in addition to breaking the ice in the group, its main objective is to convey the need to practice the observation, to look at the details, to feel comfortable observing and above all to be observed by a person who does not know us. This is a very important aspect to be able to master the technique of empathetic interviewing. It is very good to share the results of the activity; share how they think about the picture. It is also good to seek the opinion of the rest of the people in the group, who although they have not drawn that person, can give their feedback on some aspect that they would have drawn. Feedback is a very useful tool, which, in addition to seeking the participation of the team and creating a relaxed atmosphere, serves to learn to listen opinions of others and be proactive. It teaches us to know how other people see things, that is, it teaches us again to be empathetic.

Figure 3.4   Watch these people carefully: What? How? Why? (Helena Lopes on Unsplash)

The second activity that we propose will serve us to practice the three big questions that we use in observation and that can also be used in the empathetic interview. In this case, the activity takes place in observation. To do this, we take a photo of several people doing some activity, or observing one of them (Figure 3.4 is just an example). We propose to make an observation in stages, starting with a superficial observation until we reach a deep or emotional observation. And for this we propose to use the three main questions: What?, How? and Why?.

This activity can be done first individually and then by teams.

Activity 2. Observing by means of the three big questions: What? How? and Why?

We show the participants a picture where there are people who are doing some activity. Following an order, we ask the participants to explain to the audience what people in the photo are doing, how they do it and why they do it. Before starting we must explain what we intend to obtain with the corresponding answers.

The answer "What?" is related to the pure observation of photography. Responding it, we will describe what the people of the photograph are doing, writing down all the details. At this time, you must be objective and not assume anything or explain anything, but narrate what you see.

Once we know "What?,", what happens in the scene, we move on to the next question: "How?," which is directly related to understanding the scene, that is, how the people in the scene do what they do. We look at whether it requires any effort, whether they seem happy, in a hurry or in pain, whether the activity they do impacts them positively or negatively. At this stage, it is recommended to write descriptive phrases, with a large number of adjectives.

Finally, we come to the most empathetic part of this activity, the question "Why?" With it we try to interpret the situation, to get into the skin of each person. Do we wonder why the people in the scene do the things they do? And why do they do it that way? At this moment, we move onto the motivation of the action, to the feeling or to the emotion that moves the subject of our observation. Now is the time when you can guess, make assumptions of the scene, based on motivation or emotions. Try to get into the scene, to project on you the situation you are observing, to live it, and then to analyze the conclusions... some of them may be unexpected!

As in activity 1, we share the story created by each person or group from the observation work. It is very good to verbalize the story of each one, and contrast it with other people, starting from the same photo. This makes them understand that there may be different points of view and that they learn to listen and ultimately work on empathy. It is very good to ask them the question of whether they managed to get into the skin of the people in the photography. Have they felt and lived that story as if they were there?

The third activity that we propose is always very groundbreaking. It is done in pairs and requires concentration, a great job of observation and sincerity. This activity makes you end up connecting totally with your partner.

Activity 3. Looking into your partner's eyes

The activity is carried out in pairs. Try to sit in front of a person, silently, looking into each other's eyes, for a limited time. In this case, we propose 4 min. During this time, many things happen: there are moments that make you want to laugh, to disconnect and look away, there will be moments when you feel tense, then more relaxed, even after a certain time you feel calmer and finally in connection with the person you look at with your eyes. Is this the empathy we are talking about? Are we feeling it?

This third activity can be done in combination with first activity of drawing your partner. The proposal is to begin with the portrait drawing and continue with looking into the other's eyes. After this intense time, it is good to move the group members to talk about their predominant feelings.

3.3.2 Interaction

The activity that we propose in this section (activity 4) can be seen as a game. In this activity, we will interact with the rest of the people participating in the course, in which we will all be on equal terms, but we will not have all the information about ourselves. The activity is explained within a box.

The result of this activity allows us to reflect on how we interact with others and how we feel based on how the rest of the participants interact with us. Thus, we learn to address others so that they feel at ease and respected. Then, we continue to work on empathy.

Activity 4. The game of greetings

The activity will be carried out with the whole group of people who attend the course. We will need for this some playing cards or similar. Once the cards are mixed, one is dealt to each participant, who will not see it and will place it stuck on the forehead. Thus, each person can see the cards assigned to the rest, but they will never know their own card.

The dynamics of the game are as follows: all participants will walk around the room, and, at each time they meet a person face to face, they will greet them based on the range they have on their card stuck on their forehead. So if you find someone from the face cards: the valet (knave or jack), the dame (lady or queen), and the roi (king), the greeting will be very ceremonious and give an elegant salutation. And to the numerical cards, a less effusive greeting is intended that will decrease until you reach number one or two, which can almost be despised.

For a limited time, we will see all participants wandering around the room and receiving greetings from the other participants.

At the end, each participant will be asked to be sorted by the social rank they believe each one has, based on the greetings perceived by the rest and based on how they felt. There will be those who have felt kings, and there will be those who feel the last vassal.

Before starting the game, the people who are going to energize it can prepare the cards, depending on the number of participants, to ensure that there is that social variability and avoid very close cards (e.g., that there are ones and twos together).

3.3.3 Interview

To practice how to do interviews, we present a series of questions that we have chosen from those proposed by the psychologist of the State University of New York, Arthur Aron to create proximity between people. In the box below, there is a selection of questions to practice the empathetic interview [35].

Activity 5. Practicing the empathetic interview

The objective is to carry out this questionnaire in pairs, in such a way that a person A asks a first question to person B, and then person B will ask that first question to person A. For the second question, we change the order of the person who begins to ask. You can agree at the beginning of the activity a number of questions that a person can refuse to answer, in the event of feeling invaded his privacy, or the couple has not achieved an empathetic climate.

These would be the proposed questions [35]:

1.   Given the choice of anybody in the world, who would you want to have dinner with?
2.   Would you like to be famous? In what way?
3.   Before making a phone call, do you ever rehearse what you're going to say? Why?
4.   What would constitute a "perfect" day for you?
5.   When did you last sing to yourself? To someone else?
6.   If you could live to the age of 90 and keep either the mind or the body of a 30-year-old for the last 60 years of your life, which one would you choose?
7.   Do you have a secret hunch about how you will die?
8.   Name three things you and your partner appear to have in common.
9.   For what in your life do you feel most grateful?
10. If you could change anything about the way you were raised, what would it be?
11. Take 4 min and tell your partner your life story in as much detail as possible.
12. If you could wake up tomorrow having gained one quality or ability, what would it be?
13. If a crystal ball could tell you the truth about yourself, your life, your future, or anything else, what would you like to know?
14. Is there something that you've dreamed of doing for a long time? Why haven't you done it?
15. What's the greatest accomplishment of your life?
16. What is your most treasured memory?
17. If you knew that in one year, you'll die suddenly, would you change anything about the way you are living now? Why?
18. Share five pieces of information about your partner that you consider positive characteristics.
19. Make three true "we" statements. For instance: "We are both in this room feeling..."
20. Complete this sentence: "I wish I had someone with whom I could share..."
21. If you were going to become a close friend with your partner, what would be important for them to know?
22. Tell your partner what you like about them, this time saying things you wouldn't typically tell a stranger.
23. Share with your partner an embarrassing moment in your life.
24. What, if anything, is too serious to be joked about?
25. If you were to die this evening with no opportunity to communicate with anybody, what would you most regret not having told someone? Why haven't you told them yet?

26. Your house, containing everything you own, catches fire. After saving your loved ones and your pets, you are able to save ONE item. What would it be? Why?
27. Share a personal problem and ask your partner's advice on how he or she might handle it. Also, ask your partner to reflect back to you how you seem to be feeling about the problem you have chosen.

The following activity (number 6) will allow us to train the team of interviewers in preparing relevant questions in order to extract people's knowledge. The duration of the activity will be 15 min. More time could lead to mental fatigue. The people who direct the activity will be able to participate depending on how the questions arise, and facilitate the deduction of the object they are discovering in this activity.

Activity 6. What is in the box?

This activity consists of discovering a hidden object that is in a box. So that it is not something intuitive, something strange to the normal work scenario is recommended. That is, if we are in a classroom, then we select an object that has nothing to do with office supplies or something that is normally in a classroom.

Each team member asks the person leading the exercise a question about one and only one feature of the object inside a closed box. Questions will be asked on a rotating basis. At the beginning of the activity, it is good to agree on a mandatory minimum number of previous questions, before any member of the team launches a possible deduction of the object. As a recommendation, it is suggested that the agreed number of questions should always be greater than twice the number of team members, to ensure that at least each participant can ask at least two questions.

Each team member should write down the questions that are asked, and the answers of the person who directs the activity. The objective of this is that they can rely on these questions to build a new one that can be more revealing than the previous question asked, and that can complement them, that is, as if the same person was asking. This is what we call the technique of concatenated questions.

3.4 Conclusion

In this first phase of DT, called empathy, the main objective is to identify what is really important about the target people, what their daily problems are, how they interact with the environment and what they need to solve. That is why the development of empathy is a key point in the DT methodology: it provides the insight needed to define the problem and to build the rest of the project.

Empathy is defined as the ability that people can have to put themselves in the place of others and feel them as our own. In order to be empathetic, we must work on a series of steps: observation, empathetic interview and immersion.

Observing people in their environment provides us insights into what they think and feel, and gives us an idea of what they need and their important problems. For this we have to ask ourselves three big questions: "What?," "How?" and "Why?" users do what they do.

With an empathetic interview we make a first contact with the target people and interact with them. Our goal in the interview is to keep the person interviewed emotionally active, seeking their maximum involvement and complicity in the interview. To successfully conduct an interview, we must take into account the different phases through which we must pass: presentation of the interviewer, presentation of the project, construction of a relationship, evoke stories, explore emotions, perform counter-questions and finally thank and close the interview.

And finally in cognitive immersion, with all the information previously collected in observation and interviews, we are already prepared to live the experience of putting ourselves in the shoes of the people that we just interviewed. In this phase, we intend to feel what the user feels.

Empathy is related to support, understanding and active listening. Some characteristics of an empathetic person are that he knows how to listen and he is sensitive and tolerant. Being empathetic has many benefits, such as increased self-esteem or emotional development. There are people with a predisposition to be empathetic, just as there are people who are leaders by nature, but there are others who need to work on it like any other social skill. Empathy can be trained and developed through active listening, respect and a series of activities and guidelines proposed along this chapter.

Chapter 4
Definition
Manuel Caeiro Rodríguez[1] and Íñigo Cuiñas[1]

This chapter guides the readers to manage the jump from observed and gathered data to the definition of the problem that should be solved. The time and efforts invested during the empathy stage must be converted into a clear, significant and motivating problem definition, as a point to support the design of a solution well adapted to the user needs. A collection of tools and exercises is also provided to guide the readers in the path of identifying and defining the underlying problem.

4.1 Introduction

Unlike other project development methodologies, design thinking (DT) does not start with a problem to be solved, given as the initial task to be addressed. The problem has to be defined first in the context of a challenge with real-world implications, considering the users concerns. Thus, the project needs to be considered from its roots, and this involves the objectives and needs based on the results of the previous empathy stage, to identify the essential problem that we are going to solve.

Basically, to solve a problem properly, first we need to define it correctly. And this is the aim of the second stage of DT, called definition. From data gathered at the empathy stage, a problem is clearly stated and the team is prepared for ideating solutions, as depicted in Figure 4.1.

Beyond the identification of the essential problem, the definition of the problem involves other two key issues as a part of the DT methodology. First, when considering the development of innovative solutions, the definition of the problem offers us new possibilities for innovation. The problem definition can provide a fresh starting point towards the provision of a solution for a new problem, rather than a new solution to an already known problem. A new view about the problem is an opportunity for innovation with fresh ideas from the DT external team, different from those enunciated by members of the company or institution that is looking for a solution. Second, after the empathy stage, when we have listened to our users and clients and have a clear idea on how they think, how they feel, how they act and what their concerns really are, we have so much information that we need to process and condensate it in a

[1] atlanTTic, Universidade de Vigo [GID DESIRE], Spain

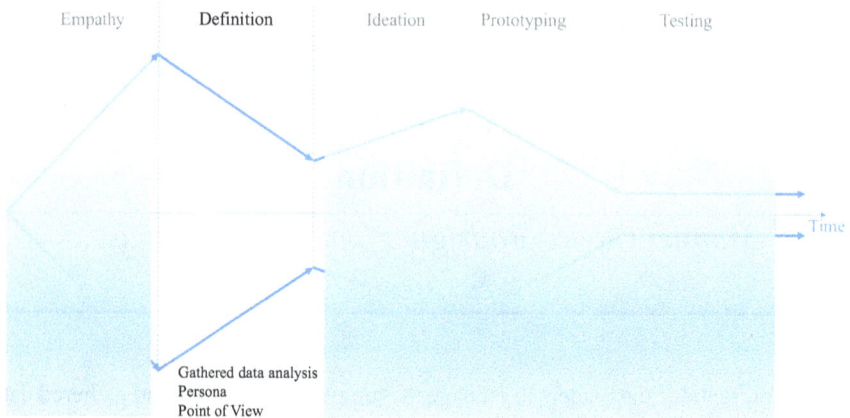

Empathy   Definition   Ideation   Prototyping   Testing

Time

Gathered data analysis
Persona
Point of View

Figure 4.1   Definition stage in the DT process

suitable way for the further stages of the DT process. Of course, the correctness of the problem diagnosis depends on how capable we were of observing and understanding our users in the empathy stage, but now we need to synthesize all that information in a more manageable construct. Therefore, we need to capture all this knowledge in a proper definition of the problem to guide us towards the achievement of suitable solutions.

At the end of definition stage, the aim is to have a good point of view (PoV) that defines the problem in a motivating way. The next section is devoted to this concept, which should inspire the team to provide ideas to solve the defined problem along the following DT stages.

There are several tools that support the process in a structured way, and they are also explained in this chapter. Processes like empathy maps, efforts and results maps, affinity maps and some others are proposed in following sections as the way to go from individual responses to group ideas that define the users/clients profiles and needing.

4.1.1   How to identify the essential problem?

The identification of the essential problem is a special key point because many times the problem that we established at the beginning is different from what the clients or companies assume that it would be their main problem. To get a better understanding on this, let's look at the following example, regarding a cosmetic company. A shampoo maker is suffering a sales crisis: its customers are not buying its product anymore. The board thinks that a new shampoo formula will help win back those lost customers, and so they assume that a new product needs to be brought to market. The company says, "we have a problem with the shampoo formula." Thus, we could define the problem as: "the formula of the shampoo may be outdated compared to the competition, so it needs to be improved." The next step would be costly for the company: they need to buy new patents, introduce new solutions, try new ingredients, etc. Each of these actions costs large amounts of money and takes a lot of time. But what if the problem is not the shampoo formula? The company must be very sure that this is the actual reason why customers stopped buying the product. In other case, the company will spend a lot of the money, although it may happen that users still do not buy their shampoo.

If we take this example and apply the process of DT, we would begin with the empathy stage observing the shampoos the customers buy at the shops, and interviewing to get answers from them through questions such as "Why did you stop buying this shampoo?" Doing this, we can realize that the obtained answers point to another problem. In this case, users answered that they still liked the formula of the product, its smell and its texture, as well as its results on the hair, but the bottle was poorly designed, and it leaks staining the entire bathtub. That was the actual reason they stopped buying the product. Then, the new definition of the problem could be: "users of our flagship shampoo need to keep their bathtubs clean when enjoying the smell and texture of their fresh and clean hairs because they do not want to make the shower up each time they use it due to the leaking bottle." Therefore, the bottle design was the real problem and instead of spending time and money on patents, ingredients, and so on, the company should focus on the bottle design. Besides, publicists could use this insight to design a campaign for launching the new bottle design of their best valued shampoo, remembering how good the shampoo is and how it has been modernized and adapted to new times.

This example shows the main function of the problem definition stage: finding the real problem based on the answers, experiences and views from customers, which we get during the empathy stage. The definition comes from the empathic research with users, not from preconceived ideas. Thus, the definition of the problem to be solved gives meaning to the effort made during the empathy stage.

4.1.2   How to define the problem properly?

To define the problem properly, we need to synthesize the significant information obtained about our audience so that in the next stage, we start the ideation stage in the right direction. A good definition of the problem is key to guide us throughout the rest of the project. We begin with a lot of information gathered during the empathy stage, and we must concentrate all its insight into a sentence defining the problem to be solved. It is like performing a convergence procedure after an expansion process carried out at the empathy stage, as depicted in Figure 1.2, to focalize the knowledge learnt.

A good definition of a problem must be meaningful. The end users and their needs should be at the front and center, ensuring the capture of the key issues involved in the problem.

A good definition of a problem must be open and focused on the need, not on the solution. At this stage of DT process, we should not worry about how we are going to solve the problem, but about identifying the problem itself. That is why the definition of the problem should not include any implication about the possible solution to it. For example, if we need to improve children's oral hygiene, perhaps a solution could come through an awareness program. However, such a possibility should not be considered at this stage. At this stage, we must deal only with the problem.

A good definition of the problem must have an adequate level of specification, neither too broad nor too specific. On the one hand, if it is too broad, we may not have enough resources to cover it and offer an effective solution. We would not have enough time, or enough people on our team to cover all aspects of the problem, or enough money to cover the costs of the project. On the other hand, if we are too specific, our impact in responding to our users' concerns could be very partial. We are probably only solving a small part of a larger real problem.

A good definition of the problem should also be actionable, offering us criteria to evaluate the eventual solutions. This element is very common in the engineering field. In order to be able to assess the possibilities of the different potential solutions, we need to be able to compare them based on a set of criteria that are in accordance with the definition of the problem. A good definition of the problem must also help us increase confidence and optimism in the design team. It must have a meaning that motivates us to solve the problem, because it seems important and significant to us and to the people we are working to. In this sense, the empathy phase should serve us not only to get informed about our audience and their problem but also to connect with them and, then, to motivate us to solve it. Being perfectly clear about what our objective is helps us to convince ourselves that this problem has and deserves a proper solution.

4.2   The outcome of the definition: the point of view

The PoV is a meaningful and actionable problem definition, which will allow us to work in the next DT stages in a goal-oriented manner. The PoV is the "light at the end of the tunnel" or the "Holy Grail" of the definition stage and aims to find meaning from the endless information collected during the empathy stage (understanding and observation).

A PoV involves rephrasing a challenge as a well-structured problem statement. To do this, we articulate a PoV combining the knowledge about the audience and their needs that we have obtained previously. At the end, it is a motivating sentence that enlightens the problem-solving process that will be the next step.

If you are aware of what the real problem is (i.e., by defining the problem, based on the outcomes of the empathy stage) and at whom the project is aimed (i.e., by defining the persona, based on the group of stakeholders found), then we are ready to define the PoV. To do it, we need to answer the following questions:

- WHO?..... (persona).
- WHAT is needed? ... (solve some problem).
- WHY?.... (insight).

4.2.1 Structure of the point of view

To articulate the point of view, we combine three elements: the user, their needs, and the information we have, into one inspiring sentence. Then, the structure of a PoV is as follows:

[Persona] needs [necessity] because [insight]

Persona: Because DT is user/human centered, the human is the central part of the process and therefore the starting point of the PoV for any specific problem. However, because PoV is intended to refocus the design challenge into a problem statement, it targets users with specific problems. Therefore, the description of the users, even if it is short, must include key personal characteristics that make them unique. Obviously, we need to clearly define the type of person for whom we are going to solve a problem.

Necessity: In addition, we must select the essential needs, those that are a priority to satisfy, and those that motivate the person. In this case, we will extract and synthesize the needs that we discover during our observations, investigations, field work and interviews. The necessity has to be captured as a verb because this moves to a dynamic action whereas a noun could be felt as something more static, and even worse, as a solution. Again, as DT intends to solve problems, the solution will be carried out to satisfy the needs of the users. However, the needs are not shown in terms of specific tangible requirements, but instead express an emotional/mental/physical state that the future solution promotes, unconsciously, through its use. Therefore, it is recommended that the "need" part start with non-continuous verbs (e.g. abstraction verbs, possession, emotion). For example, I do not need a car to go to my work, what I really need is to enjoy the commute from my home to my work: as this is an activity I should do every day, having some kind of satisfaction is important to be motivated in order to find the best mobility solution.

Insight: Finally, we need to express the perceptions associated with the needs that motivate them. These insights will not be the reasons for each need, but rather insights that we can leverage to enrich our solutions. The PoV ends with the proposal of a supposed "surprising perception." In general, if a person has a PoV, it means that some justifications must be given to support the proposed opinion. In the case of the PoV of the DT, the insight is seen as a statement that reflects a clear and meaningful perception of human behavior in a particular context. In other words, it is a justification made to support the needs of the user that is based on adductive reasoning that intends to propose the best (hypothetical) assumption or the reason why an observation is taking place.

Table 4.1 collects some PoV examples. The user must represent someone of flesh and blood person and not a typical or generic user. He/she must have the characteristics of one of the members of the interest groups that we have identified after analysing the information captured in the empathy stage. The need must always be given by a verb, since this allows us to abstract ourselves from specific solutions. In the third column, we can clearly see what those perceptions are like that will guide us

Table 4.1   Examples of PoV

User	Needs to	Insight
College student living far from campus	Go to campus 1–2 times a day and return to his home using clean transportation	It is important for her to think and live ecologically
An elderly man who lives alone in a rural area	Do physical exercise	He prefers to remain independent than having to live in a seniors' residence
An amateur beekeeper	Approach his hives	He wants to collect honey and take care of his hives

towards the solution. In the case of the elderly man who lives in a rural area, it was observed that the underlying motivation of older people to stay active is not something as abstract for them as being healthy or fit, but rather to guarantee their independence. In the same way, the beekeeper does not want a suit so that the bees do not sting him (that would be the apparent need), but to catch the honey without being punished. This will guide the design of the solution (e.g. a suit) to ensure its ergonomics taking into account the precise movements that are made when collecting honey from the hives.

4.2.2   How to find the PoV?

Along the process to find the PoV, we must consider the following issues, which can be linked to tools explained in the following section:

Specific user: In general, it is linked to a persona, that is, a fictional character whose profile reflects the characteristics of an existing group. This comes from the analysis of the observation made during the empathy phase and the state during the definition phase. Thus, the persona should reflect features taken from this group that, for example, are related to socioeconomic/demographic factors, needs, wishes/hope, or cultural background.

The needs of the user: What the user needs can be seen through the empathy map, the affinity map or the relationships that have been found between the observations and the construction of thoughts of the conceptual map. Also, this can come from the observations and discoveries of the group.

Surprising insights: To find a perception we can use deductive reasoning like detectives do. If the assumption that comes from the observations is known (it can be the result of the affinity map or the concept map or simply what was defined as the user's need), it is possible to interpret or answer each one of the following questions: "WHY?" "WHAT?." If the answer is short, or if it is not satisfactory in the way it is formulated, use the *five whys* technique (cf. Section 4.4) to get to the bottom of the matter.

The PoV will be a reformulation of the insights identified according the six criteria below. Thus, a PoV should be:

Authentic: It must be born from your own personal analysis of the experiences. This means, the PoV is the result of the work performed by you and your team during the empathetic interviews, and it is not a consequence of preconceived ideas provided by others (your boss or your contractor).

Not obvious: Not simplified, presenting itself as a simple caricature. We are not looking for a simplification of the problem, or for a small piece of problem. At this point, we have to focus on defining the full problem. When dealing with solutions, it will be time to limit, if it is the case, to partial solutions, or just mitigations, but not at this time.

Informative: Make it revealing when read. It should provide meaning to things. The PoV should be descriptive for any reader, even for those that were not involved in the empathy stage. Nothing must be assumed as known by the reader, and PoV has to be self-contained.

Revealing: It should provide information about human behavior in a given context.

Inspiring: As the PoV is also the starting point of the ideation stage, its formulation has to promote creative thinking. The writing should be motivating, boosting you to an active response.

Memorable: Concise and easy to remember. This is important as the creative team must have in mind the PoV during all time they are providing ideas to solve the problem.

The PoV is like the end point of a long and difficult process of empathy, and at the same time, it is the starting point of the solution that is going to be carried out in the next DT stages. Finally, it is difficult but very important to end PoV even if you are not sure it is correct. It is necessary to continue advancing in the DT process, taking into account that, as an iterative process, we can return to the definition stage at any time, when needed.

4.2.3 *Individual work and group sharing*

Keep in mind that a PoV requires reflection, so it is a good idea to search for it individually. Also, organizing your information from the empathy stage by means of pieces of paper or post-it notes would help you in moving from the world of the ideas (your mind) to more tangible elements. A good strategy to integrate your search within teamwork is to organize group sessions where each member presents his/her PoV to the whole group explaining it in a detailed way, all participants paying attention by active listening, and then discussing together towards a definitive PoV for the project is agreed upon.

As indicated, the PoV is the goal of the long and laborious process of empathy-definition, and, at the same time, the starting point for the creation of a new solution. During the time invested in this task, you should take in mind that this is a really important stage in the DT process. Thus, it is not worth saving time or resources by formulating the PoV in a shallow way. A badly formulated or too simple PoV would probably lead to a bad or simple solution, which could not satisfy users' needs. A PoV can be laborious and difficult to formulate, but it is essential to move to the next DT stage. It is important not to give up until you have found it.

4.3   Definition tools

The definition of the problem is greatly facilitated when we identify the different stakeholders related to it, and then we classify the users according to these groups. Our goal is actual people, and the identification of groups of interest helps us to characterize these people better and to keep away from the more typical users. It is worth noting that the stakeholder identification is made after the empathy stage, once a collection of users has been observed and interviewed: this is not an a priori classification. After the identification of the groups of interest, the Empathy Map will be drawn up, which will allow to capture the ideas that the team of interviewers has obtained from the answers noted to be grouped in an organized manner. This tool is especially useful to organize the gathered information. Efforts and Results Chart is another tool that can be used to capture the empathy results. Then, the Affinity Map is a good tool to arrange ideas and to find key associations.

All these tools (empathy map, efforts and results chart, and affinity map) are supporting methods to arrange the knowledge obtained at the empathy stage and to unlink that from specific individuals to reach insight applicable to the collective of target users. Finally, we elaborate the PoV, which will shape the definition of the problem based on the *five whys* technique.

Along all the process, the proposal is to manage physical written ideas (in post-it notes or pieces of paper) because as Jon Kolko indicates "we must put data out of the cognitive sphere (our mind) and the digital sphere (our computer), to make them tangible in the physical world (the wall, a paper fixed on that) with a coherent visual structure. By doing that, we liberate the memories from natural limitations of our brain and from the artificial organization of technology. It means that ideas are no longer connected to a specific person or situation. Thus, we can freely move and manipulate the contents, and all data set can be observed at the same time. This allows us to concentrate on the facts and feelings, and not on who generated these facts or who felt these emotions. Then, we discover the implicit and hidden significance, connecting pieces of data among them that in other way they were unconnected" [36]. This means that when our mind has been liberated from the task of remembering ideas, data and feelings, it could focus on discovering the connections among all the pieces of information gathered. This process opens the door to a more creative way of analyzing and organizing the collected information, and it is a constant along the definition stage.

The following paragraphs introduce those different tools, most of them integrated in that philosophy of putting the ideas in physical support for liberating our brains from memories, and for being then free to manipulate all the information, creating new and inspiring connections.

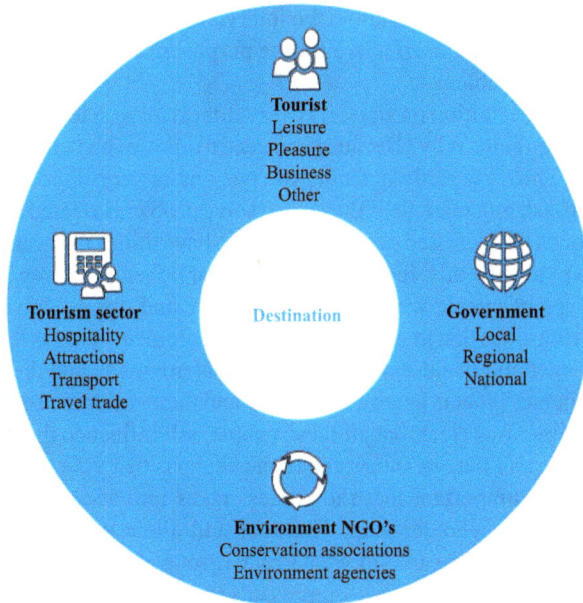

Figure 4.2   An example of stakeholders' map in a touristic destination

4.3.1   Stakeholders and persona

Taking into account the information obtained during the Empathy stage and our observations, a good strategy is to define stakeholders among all the agents involved. For
example, in a touristic destination, different groups of people can be established
based on their interests in the place. Thus, we could identify the government interesting in promoting its places, the tourism sector looking for attracting clients and doing
business, the NGO's wanting the preservation and protection of the environment and
historic places and buildings, and the tourists expecting some days of fun, action and
cultural visits. A correct classification of the people involved in the project in groups
of interest will also help us to better organize our work and to better allocate our time
(cf. Figure 4.2).

Once we have defined the stakeholders map, we have a visual idea of the distribution of the interviewed people. At this point, we could realize that some of the groups
are not conveniently defined, or that some of them were underrepresented during the
empathy stage. If necessary, we can re-interview users to complete the information
we have about them from the perspective of the group of interest to which they belong.
For example, we can prepare a specific interview for each group with the aim of con-
firming their needs, detecting unexpected attitudes of their agents, and characterizing
situations not initially foreseen that allow broadening the definition of the problem. In
this case, we will introduce new questions incorporating all the relevant ideas that can
be considered within the various groups of interest. These ideas will be structured by

fundamental areas or themes that allow the interview to be oriented in a comfortable sense for the interviewee, and effective for the purposes of capturing relevant data in the definition of the problem.

In the problem definition process, it is very important to select the right group of users once we know them. Why? Because you need to know who you are going to solve the problem for. Thus, correctly defining the persona is very important. Go further and, for example, ask other people if they try to solve problems related to playgrounds. Consider all groups of interested people, those who have things in common or who are influential for a place like this, for example, mothers, fathers, children, grandparents, and any other type of people who come to this place and use it. However, there are surely many other groups that may be affected by this problem, such as the city council, since it is the owner and decides on the changes that are made and the money that is invested, or the nursery that is nearby from there and their children use that park every day at the same time. It is clear that all these people will influence the decision that is made for this place and may be interested in the changes that your project will make. Therefore, it is very important that the process takes into account all stakeholders and puts them on a stakeholder map. This tool will allow you to recognize and/or consciously decide for whom and for what group (group=person or organization that, with a set of characteristics and needs, represents this particular set) you are going to solve the problem you have defined, that is, for who you will create the solution. You must add characteristics about your persona, such as age range (for example, children between 5 and 12 years old), nature, needs, financial wealth, tastes, and restrictions. Sometimes some aspects such as nationality or gender can also be important.

Figure 4.3 depicts two examples of personas. Both examples are related to travelers (in an airport or in a hotel), but after the empathy stage, the DT team was able to identify different stakeholders in a group that could be simplified as "travelers."

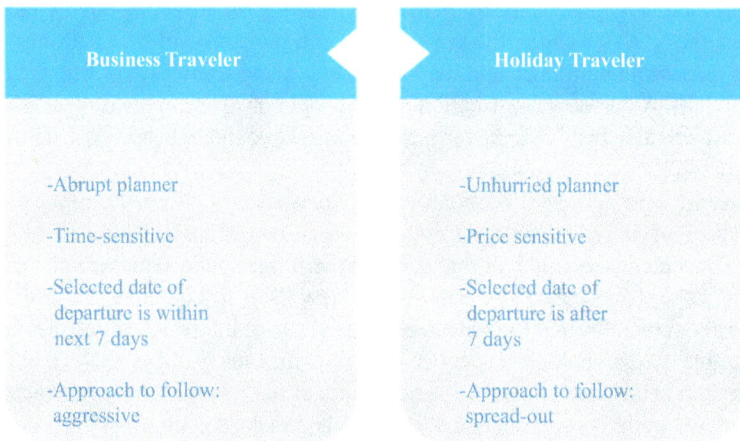

Figure 4.3   Persona examples

Some details make important differences: the way for planning the trip, the main sensitivity to decide the destination or the mode of transportation, the anticipation in selecting the dates, or even the temperament during the trip. All these are represented in the two clearly different examples of persona in Figure 4.3.

4.3.2 Empathy map

The Empathy Map is a tool that facilitates the organization of the ideas obtained from the analysis of the verbal and non-verbal languages of the interviewed users. It helps us to better understand people through a deeper knowledge of them, their environment and their unique vision of the world and their own needs.

We can prepare the empathy map from a large sheet of paper, divided into four sectors, which is glued to a wall for greater convenience (cf. Figure 4.4). Each sector represents a collection of ideas obtained by the team from the responses of the users. For a better interaction between team members, each sector offers the following perceptions, which in the team's opinion the real user has manifested in the scenario of interest:

- What does he/she think and feel? What tensions between the users can we capture? What contradictions do we see? What are their concerns? What are their aspirations? How are they surprised?

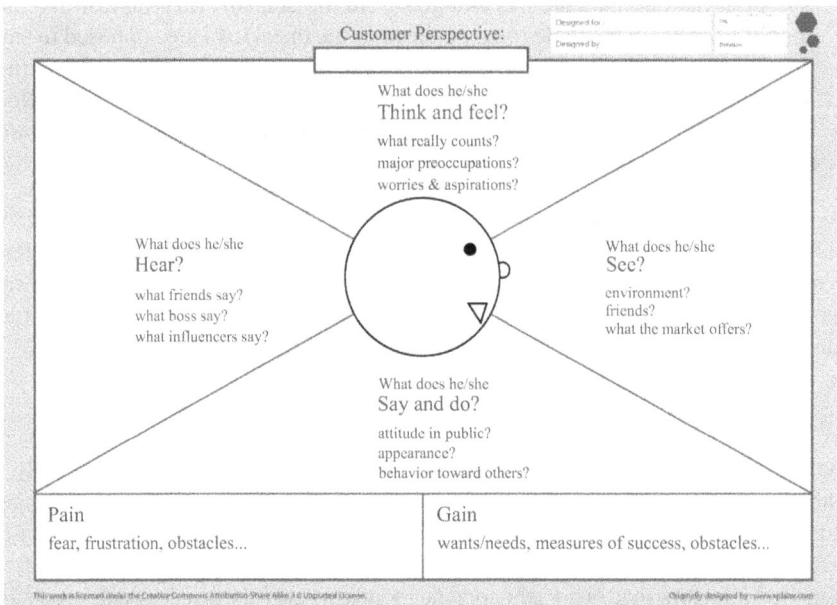

Figure 4.4   The empathy map [37]

- What does he/she see? What does the environment look like? Where do people work or study? What audio–visual information is there? How do they interact with each other (friends)? How is the body language?
- What does he/she say and do? How is the attitude in public? What messages do they convey? How do they talk to each other?
- What does he/she hear? What messages do they hear? How do they access knowledge and/or information? Where do the challenges come from?

The dynamics that the team must follow to build the empathy map in the definition scenario is as follows:

- All team members must stand near the empathy map. Each idea, in a very few words, is written on a post-it note and stuck on one of the four sectors of the empathy map. At this stage, the only rule would be to read out loud what was written so that the whole team knows what the others are perceiving and not to repeat the same ideas. This process should be dynamic, but respecting that only one participant is reading one idea at each moment and only one post-it is stuck on at a time.
- All ideas must be original, and the team must strive to maximize the possibilities of data extraction offered by the interviews.
- No one should make judgments about the ideas of others. There are no leaders who point to the absolute truth, and all ideas are equally valid. There are not stupid or secondary ideas or knowledge, as even the smallest detail could be interesting for the problem definition.
- It is essential that the comings and goings to the empathy map be fast. No one should sit down at the table to agree on what the interviews are supposed to say.
- There is no consensus process. Concepts must flow intensely. There will be time and methods to refine, and future DT stages to look for ideas and solutions. Now the advantage is that everything is raw, and there is total freedom to delve into the data obtained from the interviews.

From that immersion will emerge the best concepts to build the definition of the problem. Once ideas have been obtained and the empathy map covered in its four sectors with notes, each and every one of the sectors of the empathy map is reviewed again, and the labels are reordered, in a process of collective refinement, in that all team members try to refine the set of ideas noted down.

4.3.3   Efforts and results chart

The empathy map can be complemented with the efforts and results chart. Basically, it is about compiling in two columns the fears, the possible frustrations, the challenges and the obstacles that our audience will have to face (this would be the column of efforts), and what our audience wants to achieve, their desires and objectives, as well as those indicators that our users consider a measure of the success achieved (the results). For this, we also use post-it notes and proceed in the same way as with the empathy map.

Table 4.2   Example of an efforts and results chart

Efforts	Results
He loves to taste local food in good places	He does not want to lose his good health
He loves to play sports and to do active plans	He keeps him fit and relieves stress
He loves to visit artistic or historic sites	He learns a lot on local culture and history

Table 4.2 shows an example of an efforts and results chart related to tourists that come to a specific destination. They have different preferences for the activities to do and also different motivations for those.

The efforts and results chart could be valuable to have additional information on what is really important for the target users, and it also provide some insight on the roots of the problem we are trying to define.

4.3.4   Affinity map

Affinity mapping is the process of grouping notes into similar topics or categories. It is about creating groups of ideas, categories or relationships among them in a way that helps us organize and simplify all the issues that we have put on the table.

The process is simple. You will need post-it notes and space to post ideas (a wall or board of some kind). It is important to have room so that everyone in the group can move freely and interact with the notes and post space. It is also important to create an environment where attendees feel open to sharing ideas. Use a marker pen so notes can be easily related or grouped together. Group post-it notes that are the same and those that are very similar. Put peripheral notes into their own categories and then continue grouping until you have a series of separate groups in place.

Figure 4.5 depicts an example of affinity map, related to a new way of managing the clients in a hotel. This map groups the advantages and disadvantages of the hotel characteristics, the possible offers (upgrades) at different services and the requirements for the personnel at reception desk.

This tool helps in moving from the results of the interviews (which were the bricks of the empathy map) to the roots of the problem we aim to define. It could be identified as a step to jump from the persons to their problems.

You may want to take pictures to remind yourself of key points after the session. The collaborative approach should create a level of enthusiasm and excitement, but it can be approached differently depending on your organization and group.

4.4   The five whys technique

A possible way to identify insight would be through the typical deductive reasoning of detectives. For this we can apply, for example, the technique of the five whys [38]. This is a technique to explore the cause-and-effect relationships underlying a particular

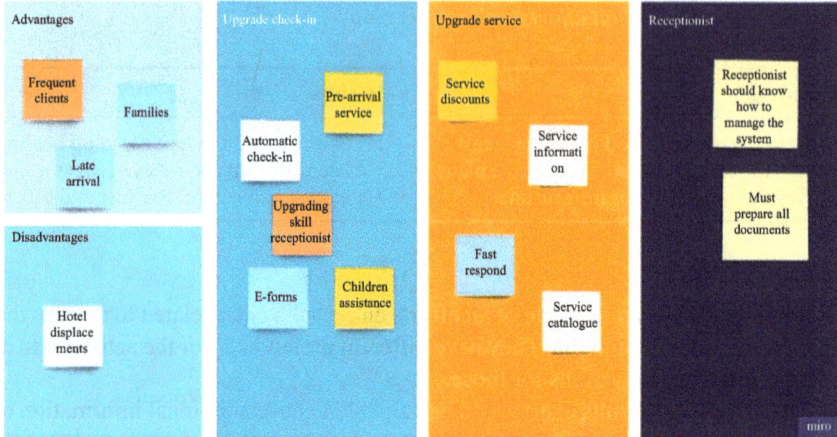

Figure 4.5   Example of an affinity map in the tourism sector

problem through iterative questions, and more specifically, repeating the question why? Each answer forms the basis for the next question. The five in the name of this technique follows from an empirical observation: the number of iterations normally required to solve the problem is five. In any case, we can continue asking ourselves the reason for the last effect in order to find a new cause as many times as necessary. An example can be observed in the sequence below.

The girl's five whys

The girl looks sick.

1.   Why? She is sick.
2.   Why? She lacks basic nutrients.
3.   Why? She eats badly.
4.   Why? She refuses healthy food.
5.   Why? The boys and girls around her always eat junk food.

This method is a good strategy to refine the PoV's phrasing, improving the value of the definition by going to the roots of the problem instead of being satisfied with the first enunciation.

4.5   How to be sure that we have defined the problem properly?

You will never be 100% sure that you have defined the problem properly, but you can do some activities to find out if you have really done it right.

First of all, spend as much time as necessary to carry out the empathy stage. Ask suitable open questions, conduct the interview according to the most appropriate rules and immerse yourself in the environment, if possible and necessary. Having as much information as possible is fundamental for constructing a good definition of the problem.

Second, build the empathy map with the information collected, as explained in previous sections. The empathy map can be made as it appears in Figure 4.4 or in another way. The most important thing is to collect all the information obtained during the empathy stage and write down what the interviewees think, how they feel, what they hear, see, say and do. Some additional information can also be added that can be grouped: efforts and results chart is a useful tool to collect these additional insights that are not easy to introduce in the empathy map.

Next, group all responses (focus on ideas, too) and put them on the affinity map, by moving post-it cards from the empathy map now grouped by topics. The names of the groups are not pre-set (the team will name them depending on the type of information it receives and put them on the empathy map). For example: infrastructures, fears, aspects related to transport, perspectives, financial issues, etc. In fact, you can be as creative as you want during this process.

You have to solve the most important problem considering which problem is the most serious for the largest number of people, that is, the most important problem which solution satisfies the largest number of customer needs. And this decision must be supported by the analysis based on the explained tools and procedures: using the newly organized information, you should write a PoV.

Then, when you have a PoV, you have to evaluate its goodness and improve it, in case you consider that it could be improved. Thus, some tips on the characteristics of a good problem definition are useful. Remember that the PoV should be focused on the people who we are trying to help, instead of focusing on user-types; wide enough to allow creativity avoiding specific methods or techniques regarding solution development; authentic and based on real-world people observations; specific enough to be manageable; non-obvious and revealing; inspiring and memorable. Besides, the definition of the problem should not take into account the technology, the monetary benefits or the specifications of the final product. These aspects are the subject of later phases of the process of building a solution.

Considering the first two DT stages in conjunction, empathy and definition, we can see that the definition of the problem is done according to a process of analysis and synthesis. Namely:

- Analyzing is essentially breaking down something complex into simpler components, and therefore easier to handle and understand. Basically, this is what we do in the empathy phase when we observe and document users and their interactions, or when we seek information about the problem, classify it, and relate it to our observations.
- Synthesizing, on the other hand, consists of creatively selecting elements from the result of our analysis to build complete ideas with them. This is what we

fundamentally do at the definition stage when we select, classify, interpret, and make sense of the results of our analysis to create a definition of the problem.

Also remember that the process of DT is not linear and iterative. If you are still not sure or if you discover that you have not carried out the right problem, you can go back to the previous phase, empathize again, get new information, facts, perceptions and redefine the problem. It is okay to fail, but the sooner you do it, the better!

Although the process of analysis and synthesis is key in the empathy and definition phases, it is not exclusive to these stages. Most likely, we will encounter situations where it is necessary to analyze a situation before synthesizing new insights, and then analyze the synthesized findings once again to produce more detailed syntheses.

As discussed above, the definition of the problem with the characteristics described eventually materializes as a PoV. Revisiting our PoV about tourism, we could think that the challenge at a touristic destination is to provide a variety of activities to the tourists, but this would be a very simplistic vision. A better definition would be achieved by referring to the PoV: a tourist traveling alone needs to do activities involving local people because this way he could know more about the destination and he would feel accompanied.

Chapter 5
Ideation
Manuel Caeiro Rodríguez[1]

The objective of the ideation stage is to generate new ideas that could be a solution to our problem. The purpose is to go beyond canonical solutions (or typical engineering recipes) and to find solutions that can satisfy the specific needs of the end users, focusing specially on the key insights we have identified in the previous definition stage. The ideation stage involves thinking out of the box, questioning assumptions, exploring the limits, and looking for disruptive solutions.

The ideation stage is structured in several more detailed steps. In the first step, the goal is to get as many ideas as possible. Instead of pursuing a single best solution from the beginning, first we must focus on generating ideas, keeping in mind mottos such as more is better, do not judge ideas, and build on the ideas of others. Then, once the ideas have been generated, as a second step a review process is carried out to address the formulation of concepts that can provide a comprehensive solution to the original problem. Finally, the last step of the ideation stage involves the selection of those solutions that are considered the most suitable to be prototyped and tested.

5.1 Introduction

In the ideation stage, we search for possible solutions to the problem satisfying the needs of end users. Sometimes we do not have a clear idea of the possible solution, but simply an intuition or a vision that can work. In other cases, we have many different options that we can consider. Whatever case, the purpose of this stage is to identify some solutions to be developed and experienced in the next two stages, namely prototyping and experimentation.

Ideation is a creative thinking process that involves the creation of new ideas and solutions. It is a fundamental moment where it is decided how to use the knowledge about a problem that arose for the user in the empathy phase and how to respond to the problem already defined in the definition stage, and for which solutions are being sought, as it is depicted at Figure 5.1.

[1]atlanTTic, Universidade de Vigo [GID DESIRE], Spain

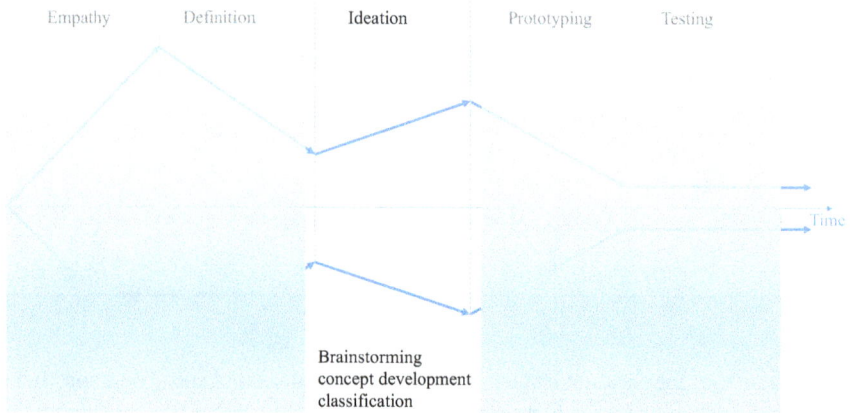

Figure 5.1   Ideation stage in the DT process. The project progresses through expansion and concretion stages. Once we have specified the problem with a Point of View (PoV, cf. Section 4.2), a new phase of creative expansion begins.

In the ideation stage, cooperation within the team is a fundamental element. There is no other time in the DT process that allows team members to cooperate with each other so closely. In addition, it is fascinating to observe the process of generating ideas when they come out of nowhere. Several ideas can appear at the same time. Ideas can be based on other ideas. Sometimes new ideas arise from a word that is pronounced or by a coincidence. Sometimes, if we speak randomly, a word can be the beginning of a chapter and, therefore, everything that comes before ceases to matter because the project has just found a new path, a new route on the map. These are some of the most valuable moments in the DT process: when a new path is found that leads to the destination point.

The work environment that helps to find the moment (the Wow! moment) is only generated when there is a mental and personal commitment from each member of the team.

In the ideation stage, we can use brainstorming [39], a very popular ideation tool. In general, brainstorming is carried out in a very free and open way, but for this stage, we propose a much more structured and disciplined version of it, trying to avoid non-productive situations and to improve results. Another key tool in this phase is concept development. Based on the results obtained in the brainstorming, this tool guides us to organize the ideas into coherent sets and present them in the form of a robust "concept." An "idea" can be collected in a post-it note, while a "concept" usually requires a panel in which several post-it notes are collected and grouped in a more or less organized way. Finally, a polling or some kind of contest is needed to select the concept to be developed in the next prototyping stage.

5.1.1 Ideas before ideation

In the previous stages, empathy and definition, as we get to know the end users in depth through empathy and understand the problem through definition, in many cases, it is inevitable that ideas of possible solutions will arise. If this happens, stop, do not think about ideas before reaching the ideation stage. The ideas that occur to us during the empathy and definition phases should be written down in a notebook where they must be forgotten. Failure on this can prevent us from giving our creativity a chance and not addressing the really important problems. Basically, we can be doomed to define solvable problems with the ideas we already have. In other words, the canonical method of solving problems by applying what we already know is not the best method to meet the challenges of a constantly evolving world.

5.2 How to ideate?

The ideation stage should start with the PoV as the key reference. The PoV defines the problem or the needs to be satisfied, identifies the end user to address the solution and, through the insight, shows the main issues to be considered in the solution.

In addition to the definition of the PoV, as it is usual in engineering projects, a set of design criteria should be considered. For example, the solution may be made available for use on mobile devices, or it should work properly even without an Internet connection, etc. Of course, criteria should not determine how the solution has to be, but rather frame it within certain constraints.

During the ideation stage, it is important to keep in mind that in the DT process, there are no simple algorithms or strictly defined tools that ensure that if we start from point A, we can get to point B. It would be nice to present the activities as if they were on a map, with information about the different routes that take us to a destination point or, sometimes, to several points that can even go beyond the limits of the map itself. Different paths may lead to a variety of solutions for a given problem. To illustrate this, we can take the example of the problem of how to transport food from one side of the river to the other where a family is living. There is no single way to solve it. Food can be transported in a wooden boat or by swimming with a waterproof bag, crossing with a rope or even creating a wooden bridge that would allow us to use a car. Moreover, we can build a steel bridge over, allowing us to drive a truck full of food or a highway for hundreds of trucks crossing every hour. We could also transport food by using a helicopter or a plane if two landing strips were built on each side of the river. All these examples show that there can be an endless number of solutions.

In the first approach, a large number of ideas and solutions are desired, some of them can be crazy ideas, but as we move forward in time and deeper in the search of a solution, we will begin to be more realistic, and ideas will show a better quality. Based on the example of transporting food across the river, it is important to consider funding, amount of food required, feature of the river, time, and the possible needs of people on both sides of the river. Thinking about this information leads us to discard many of the solutions we have considered and perhaps to consider new ones. We could think that the solution is that the family could settle on the other side of the river and, therefore, there would be no transport need.

5.2.1 What is important to ideate?

To avoid the possible problems of the Ideation stage, it is necessary to look at technical and physiological aspects that guarantee comfortable and efficient work. For this, there is a list of requirements that must be considered, but this list is not closed since each team should adapt it to its particular project.

Three important requirements should be taken into account for the ideation stage:

- The end user (persona), for whom the solution is being developed. We must refer to the information collected about the end user in the Empathy stage. Never forget that our ideas are intended to solve a specific problem, changing the life of a particular user.
- The problem, as it has already been defined by the team in the Definition stage, synthesized in the PoV.
- The whole DT process. The Ideation stage is a part that should include and refer to all the information that has been obtained previously and to the activities that are going to be carried out in the next stages.

5.2.2 Technical aspects

An appropriate workplace is needed to ideate. Its size is important, as it has to be adapted to the team and the type of project. Access to fresh air and comfortable furniture should be provided.

Natural or artificial light is also required. It is desirable that the ideation space has natural light, but if you work late, you will also need an artificial light source. It should not tire your eyes or interfere with your concentration (e.g., by making noises or blinking, which can be very annoying).

The ideation place should be equipped with technical support materials such as bulletin boards, blackboards, markers, pens and pencils, sticky notes of different colors, sizes and shapes, and tape. Elements that promote inspiration are an interesting complement, such as photographs, colorful magazines, small devices or things that do not have much to do with the project but that may help to find solutions to a specific problem.

Finally, access to a social area and an equipped kitchen would provide a most enriching ideation experience. After all, thinking consumes a lot of energy, so the equipment needs to compensate for that.

5.2.3 Psychological aspects

Ideation shall be organized in a way that tensions and stress among team members are avoided. Comfort to all team members shall be provided by establish agreed upon working rules (e.g., equal access when speaking shall be guaranteed). Introduce the tools that the team will use and their usage or sharing rules.

Create a schedule that is not interrupted but is flexible to carry out the creative process. For example, during the process, creativity cannot be interrupted and, sometimes, it would be necessary to increase the time devoted to a stage. In other occasions, the time allocated to a specific stage will be reduced because, for example, the team is already tired. To achieve a higher level of comfort and efficiency, adapt the individual role of each team member according to their characteristics.

5.3   How to promote the generation of ideas?

Sometimes it seems that ideation is one of the easiest and simplest things, the most creative and sometimes even an insignificant part of the puzzle. However, it is not what it seems. To perform ideation, the team has to be very creative and not have certain restrictions on their work. In addition, it is necessary to build a specific atmosphere that can encourage both individual work and teamwork, guaranteeing the freedom to create solutions. It is also important that the project team feels that there are no limits in the creation process and that even the most exorbitant idea can be real and essential to the process in which we try to find a solution to the problem.

5.3.1   Think out of the box

To generate truly creative ideas, it is very important to start by considering new possibilities. On many occasions, restrictions are considered before possibilities, and sometimes unnecessary restrictions that are not really part of the problem at hand. The exercise of "thinking beyond the limits" shows us how, on top of the specification of an apparently simple problem, we introduce new restrictions that complicate its resolution.

An exercise such as the lines and dots puzzle consists of drawing lines that go through a set of nine circles arranged in a 3×3 grid (cf. Figure 5.2). In a first step, with four straight lines, drawn without lifting the pencil from the paper, it should be possible to go through all nine points with no problem. In a second step, we must be able to do it with only three straight lines, and, in a last step, the most difficult, with just one line. The key to the solution is to think "beyond out of the box," in this case literally: "out of the box implicitly delimited by the outer dots." Such a box is a restriction that we set to ourselves, but in this case, it must be removed to achieve a valid solution. Moreover, when we are faced with solving the more difficult problem, with just one line, other kinds of "boxes" also appear, despite not so evident because they cannot be represented as a physical box, and have to be removed.

Similarly, when we try to solve problems in the technical field, we also tend to consider constraints that limit or condition possible solutions. This is fatal for creativity and disruptive thinking. If we start by accepting all the things that do not allow us to do better, the solutions we can come up with will inevitably be very similar to what we already have today. At the initial step of ideation, it is preferable to ignore constraints. Thus, we will be able to achieve new solutions. Of course, such solutions may not be feasible, but that should not be the concern at the initial step. Moreover,

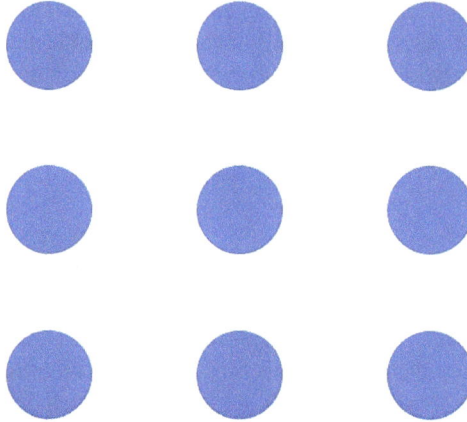

Figure 5.2   Print this figure on a piece of paper. It is easy to get through all the dots without lifting your pencil from the paper with four straight lines, but can you get through all the dots with three straight lines? and with one?

from a technical perspective, the fact that current technology does not allow a solution to be put into practice and does not prevent technological evolution from making it feasible in the more or less near future. Once we have a new idea, we can start to be creative and think about how to get rid of the constraints we have ignored.

Ignoring unnecessary limitations that we consider essential is not a skill that we can get easily and may require a great effort and practice.

5.3.2   Lateral and combined thinking. Cross-pollinating ideas

This type of thinking consists of thinking about solutions that we know to problems different from ours, but that we can relate as a metaphor or analogy with ours. The best ideas, those that have the greatest impact on people and endure over time, even resisting technological advances, are usually transferable. In many cases, a certain innovation in a certain field can be exported to transform another different field. In any case, the greatest impact is achieved when the ideas, in addition to being interdisciplinary in their application, are also interdisciplinary in their origin. Here are a couple of examples:

- Pharmaceutical company Pfizer had a line of products to help people quit smoking. However, they realized that only with medicines the real success they had was very little, so they looked at the gyms where people who want to get in shape go. Combining their products with exercise-based group support practices, their results improved considerably.
- The *accordion house*. To solve the need for space and different rooms that small houses pose, the use of mobile walls that open and collect like the bellows of an accordion is proposed. This way, mobile walls [40] were able to transform small

houses with small rooms into small houses with big rooms, despite not all the rooms are big at the same time.

This type of thinking also helps us to explore different technological scenarios and trends. Most of the productive sectors are undergoing changes derived from the introduction of technology, and these advances in certain sectors can lead us to propose new ideas in others. An example of this would be the so-called *uberization* of some sectors [41], where ideas from the collaborative economy are applied to resources that we have but that we do not use all the time.

The silly cow

This activity may seem a bit silly, but it encourages people to think creatively outside the box. Its power is precisely in its simplicity.

The objective of the exercise is to prepare three ideas or business models based on a cow. Participants will need a pad of post-it notes and colored pencils or markers.

The activity is organized in three steps:

1. We ask participants to think of associations with the cow concept and the typical characteristics of a cow (e.g., does moo, eats grass, provides milk, has spots, etc.). They are given a time of 2 min.
2. Next, we ask the participants to come up with three original ideas about the cow business. It does not matter how crazy those business ideas are. They should draw each idea on a separate post-it note. We prefer simple graphic presentations to words. Better visualizations of ideas than textual descriptions of them. Time: 3 min.
3. Time to defend business ideas. Everyone has to have the opportunity to show their ideas and explanations. We may allocate more or less time to present ideas depending on the time available.

This activity is usually quite common in courses or workshops on innovation in the business world. It allows us to see things differently, from another perspective.

5.3.3 Question assumptions

This is about questioning assumptions that we take firmly and that determine the solutions that are adopted. For example, successful low-cost airlines proposed scenarios that were believed difficult for users to accept: secondary airports, non-conventional schedules, on-board payment services, etc. However, despite all these assumptions, these companies have succeeded, and their model has been taken over by traditional companies in the airline sector [42].

5.3.4 Explore the extremes

Extreme scenarios often generate new thoughts and ideas, sometimes impossible, but that can guide us towards new points of view. For example, there are countries in which, in addition to food delivery services, there are also ingredients and recipe delivery services, so that it is oneself who prepares the food, although with help to make the purchase and specific proposals for dishes, for example to support you following a certain diet [43]. This may sound a bit strange to you, but it is something that is working in countries like Sweden or Japan.

Linas Matkasse

This is a very popular online service in Sweden. Those who subscribe to Linas Matkasse receive a delivery to their home that consists of a box of food to cook a specific meal for a certain number of people. You can choose to receive a delivery every week, every 2 weeks or once a month. They offer a variety of food boxes for different dietary options and for a very wide range of tastes and needs. There are offers for 2, 4 or 6 people, proposals for children, gluten-free or lactose-free meals, vegetarian bags, etc. For example, the Linas Originalkasse box offers familiar flavors and delicious ingredients for people who like classic dishes, but they want to try flavors that they may not have tried before. This box is also offered as an ecological alternative.

5.3.5   Change who does what

Many times innovations come as a result of changing the key roles in the value chain. A well-known example of this kind of change is IKEA, where customers were transformed into furniture assemblers. Another example can be observed in self-service restaurants, where the staff serving customers directly at the tables is dispensed with, or online banking, where customers become their own bank managers [44].

5.3.6   Inspiring questions

A good way to promote the generation of ideas is to ask ourselves motivating questions about the problem we want to solve. For example, suppose a family is designing their new house. Some questions that can be asked are: how could we unify all the family requirements in just one place? What will the family space be like in 20 years? We can also consider being someone different from ourselves, assuming a different vision of things. For example, we can think that we are Steve Jobs, or Lara Croft. From that position, think about the solution that we would give to the problem that we are addressing.

5.4   Ideation tools

Below we describe some tools that can be used for idea generation, classification and selection. They would be also useful for providing training on DT.

5.4.1 Active brainstorming

Brainstorming is a group work tool that facilitates the emergence of new ideas on a given topic or problem in a relaxed environment. Basically, it is about generating as many ideas as possible, as crazy as they may seem, as quickly as possible. The quantity is very important.

This is one of the most popular resources for brainstorming. Nevertheless, brainstorming sessions should be performed in a specific way, probably different from the mainstream brainstorming concept. The session should begin with the explanation of the rules and the familiarization with the workplace. Make sure that team members stand up and each participant has own material, post-it notes and something to write on. Standing up is important for the development of ideas, as it activates team members and provides more ideas with greater quality. The position will provide better blood flow and better body communication when explaining ideas to other members of the group. In addition, when we are standing up we get tired faster, and this moves us to conclude the brainstorming on a certain time, not keeping it for too long.

The development of brainstorming has an eminently group character. We can all come up with ideas individually but doing it as a group makes it a more effective and relevant process. Nevertheless, as in any group activity, it is important to approach it from an initial individual work. It is key to have individual time for each member of the group to think and reflect internally on the task. After, we need to establish a group time to allow each participant sharing his/her ideas and discuss them with the rest of the group members. Underlying this approach is a basic premise of group work: everyone contribution is valuable. Therefore, to facilitate good results in brainstorming, we must offer the opportunity for everyone to share and discuss.

There are certain rules for a good brainstorming session:

- The right number of people. The maximum number of people that could be involved in a brainstorm is 12. If there are more people, a good approach might be to divide the group into several smaller groups. On the one hand, groups should not be very small, since then the opportunities for collaboration are reduced. On the other hand, groups should also not be too large, because this would extend the duration of the process too much. In any case, brainstorming can be opened up to people outside of the development team. These people can help improve results by providing a more diverse view.
- The right approach. The challenge to be addressed must be clear. The PoV plays a key role in this aspect since it allows us to focus the objective of the brainstorming in a synthetic way. Particularly, it is important to make the PoV visually present in a clear way, for example, writing it in large letters on a blackboard or even on cardboard. If possible, it is also good to promote empathy. Having evocative images of the stories that have caught the most attention, or photos of the problem you are trying to solve, or the people you are trying to help would contribute to have the right approach and help improve idea generation. Use catalytic stories. For example, the case of a person who makes a mess with his medicines can be understood quite well, but if the story is about the father of a friend and he makes

us part of all the problems that this causes, we will probably feel more involved and awaken, with a greater empathy.

- The right frame of mind. Successful brainstorming requires an active, participatory, and positive frame of mind. It is important that the participants are committed to the activity and the objective. It is a good practice to do an activation exercise at the beginning of the session. This exercise involves all the participants moving, acting, and interacting in some kind of funny game. This is very recommendable, since in addition to activating blood circulation and improving brain activity, it helps to disinhibit the participants. For example, having a rock-paper-scissors contest involving all participants is often very effective. First, each member competes with another for the best of three wins. Once the initial duels are resolved, each winner looks for another winner to play with, and the loser supports their respective winner by chanting their name and encouraging them to win. The necessary rounds are repeated until only two players left, each one having their respective supporters. After this final game, the least important thing is who wins or who loses. All the participants will have a completely relaxed and participative attitude.
- The right attitude. Related to the state of mind, an open, polite and positive attitude should be shown. There are some key issues: (i) any criticism or prior evaluation is prohibited, both to others and to ourselves, since every time we self-censor we associate in our mind "idea = error," putting barriers to ourselves; (ii) quantity versus quality, every idea is welcome, and the more ideas we generate, the more we will open our minds to generate new ideas, and seemingly absurd ideas can help us come up with brilliant ideas; and (iii) seek association and development of ideas, building new ideas on the ideas of others, combining and mixing them. There are some phrases that collect these issues in a synthetic and clear way: "Do not judge the ideas of others," "All ideas are good," "There are no stupid ideas," "Only one speaks at a time," "Listen to others," "Built on the ideas of others," and "Have fun". It is a good practice to make some posters with these slogans and paste them around the space where we brainstorm ideas, to reinforce this open and positive attitude.

One object, thirty uses

This is an activity that shows the power of group ideation. The objective of the exercise is to prepare as a team a list of at least thirty possible ways of using a common object (bucket, brick, shoe, etc.). The activity generally proceeds in three stages.

1. At the beginning, the participants are asked to give simple examples of using the object. For example, in the case of a bucket they could be containing water, making sandcastles, cleaning the floor, improvised case, toy ship, etc.
2. After a few seconds or minutes, participants often run out of ideas.
3. In the last phase, the aim is to overcome participants' own mental barriers that limit their vision of objects in accordance with the most common expectations.

Participants are encouraged to provide more ideas by trying to change the characteristics of the object (e.g. its dimensions, color or material it is made of). New uses will surely appear. In the case of the bucket, we have that a cube with many holes can act as a watering can, painted with white and red stripes it can act as a traffic separator, a fountain can be created from several cubes, etc.

It must be remembered that there are no good or bad ideas in this exercise. If any group gets stuck in phase 2, try to encourage them to think of possible ways to use something common in extraordinary situations, such as on a desert island or as the basis for a gift to a loved one.

Note that some people are not very supportive of brainstorming sessions, since there are cases in which the experiences obtained in the application of this type of session have not been good at all. In fact, several undesirable situations can occur that we must try to avoid:

- Overly active participants vs. passive participants. In some cases, participants assume two types of extreme participation in brainstormings, both of which are undesirable. On the one hand, there are those who have many ideas and do not stop talking, proposing new questions that in many cases have little to do with the problem to be solved. On the other hand, there are those who say nothing and just listen. Perhaps the worst thing about these people is not that they do not say anything, but that they do not think about anything related to the object of the session. As a result, they completely nullify their possible contributions and also their "empathy" with the work.
- To each his own. Sometimes in the brainstorming sessions each participant assumes a certain position and dedicates himself to defend it tooth and nail, without considering other options and denying any other possibility. Many times, in the case of companies, this situation becomes more serious since it is combined with the position or department of the person. It occurs as a typecasting that makes communication between the team very difficult.
- Lack of control in coordination and time. In some cases, brainstorming sessions take place without any type of control over participation. Each participant intervenes at will and contributions ideas freely. The duration of the session is also not limited. As a result, imbalances usually occur in the participation of different members and sessions can go on for hours and hours with no apparent progress.
- New ideas are not feasible. There are times when the new ideas that arise during brainstorming are not feasible with the means available to the team or the company. This tends to condition some people, who tend to filter their ideas based on the conditions that they consider probable for their development. However, as will be seen in the following stages of the DT methodology, there are techniques that allow us to develop an idea and test it with end users without having to build a complete solution. Prototyping allows you to test most ideas. Therefore, considerations about the feasibility of ideas should be avoided.

The blue cards

This activity generally proceeds along three stages.

This exercise was originally developed by Stan Gryskiewicz, co-founder of the Center for Creative Leadership [45]. It is about organizing the brainstorming session in three rounds:

- First round:
  - 3 min. Each member of the group works individually, in silence, trying to find ideas. Each idea is annotated on a different (blue) card.
  - 5 min. Ideas are shared with the group. Each participant introduces his/her ideas in turns. No one comment others' ideas, just listen.
- Second round, participants are invited to propose new ideas or ideas based on the ones already present in the previous round:
  - 3 min. Each member generates new ideas individually again. Write down new ideas on cards.
  - 5 min. Share the ideas again as a group.
- Third round:
  - Continue working as a group. Now you can propose new ideas in a more open and collaborative way. Everybody can create and share dynamically.

In general, after the first round, many ideas that were already known come to light, so it is crucial to give way to a second round in which new ideas are generated from the "echo" and feedback of existing ideas. This gives participants the opportunity to reflect on what has happened and generate new ideas, perhaps by combining existing ones. In addition, this way it helps everyone to participate. Incidentally, the blue color of the cards is not important at all, the key is in the development of the activity, combining individual and collaborative work and limiting the time slots to perform each round.

5.4.2   Mind maps

Initially proposed by Tony Buzan in 1974 [46], mind maps are widely used both in the field of education and business to represent ideas or concepts related to each other by means of a keyword or central idea.

Mind maps are built around a key idea that can be a word, sentence or short text. This idea is placed in the center of a sheet of paper or a board. Taking this central idea as a reference, other ideas are added in a clockwise direction. Then, ideas are linked to the central one using lines.

According to its creator, there are 10 key elements to consider when developing mind maps:

1. A blank sheet with an horizontal layout must be used. It facilitates the overview and the arrangement of ideas.

2.   The central idea must be represented by means of a drawing or color image. Pictures often provide more information than words, and they contribute to keep the map concise.
3.   As far as possible, pictures should also be sued for the rest of the ideas.
4.   Uppercase letters are used to write down keywords.
5.   Each concept must have its own branch within the map. This helps exercise creative memory by preventing the brain from reading sentences.
6.   Branches flow and become thinner as moving away from the center, like the branches of a tree or the neural networks in the brain.
7.   The size of the branches should be compensated.
8.   Colors should be used as far as possible.
9.   The relationships among concepts and ideas with arrows and lines should be highlighted with arrows.
10.   Empty or white spaces shall be utilized to provide clarity.

Basically, to create a mind map, we place the central idea in the middle of a blank sheet of paper. Next, we jot down individual ideas around this central idea, placing them in whatever order and arrangement we wish, using colors and drawings. Finally, we join all the elements with lines and arrows. With this, we organize the map and provide a hierarchy to the information contained in it, which in turn will allow us to assimilate and memorize its content more easily. There are several computer applications available to support mental mapping [47,48].

Ultimately, mind maps help organize our memory and improve our ability to retain information and use memory more effectively. As visual instruments, they are very convenient for transmitting a large amount of information in a concise way, while facilitating its understanding. They also help us organize our thoughts and express them clearly and effectively.

Figure 5.3 depicts an example of a mind map. After brainstorming for a new attractive drink for adults, ideas such as "sweet drink," "carbonated drink," "fruit-flavored," "alcoholic drink," "alcohol-free drink," "energy drink," "glass container," "brick-type container," "attractive container," "non-refrigerated," "cheap," "nutritious," "dairy drink," "very exclusive," "portable," etc. From them, the mental map of the figure was elaborated.

5.4.3   Concept development

The ideas generated in the brainstorming process, especially in the conditions indicated above, are usually not elaborated enough to show them to potential end users or clients and expect their positive validation. Before doing so, it is usually necessary to refine these ideas. The concept development activity consists of choosing a set of ideas (e.g., from those generated in a brainstorming) and combining them to build a concrete solution that can be presented and validated. This is a kind of consolidation activity where ideas are evaluated, compared, ranked, clustered, and even discarded to achieve a set of great concepts to work on. Our goal is to achieve the concept of a product or service that meets the needs captured in the PoV.

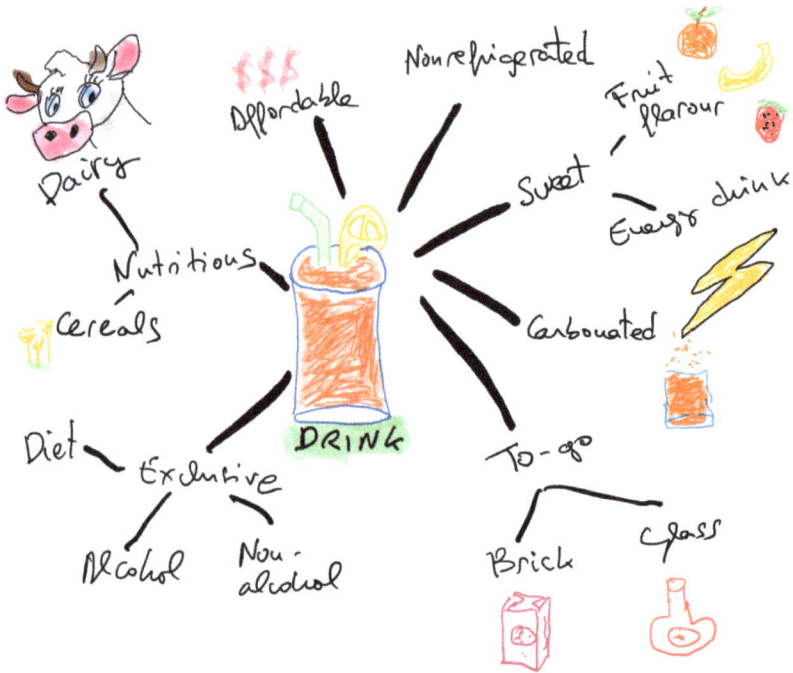

Figure 5.3   Example of a mind map

As a first step, we need to build a concept that may be of interest for the end users. This involves grouping and organizing the ideas, considering issues such as their viability or their degree of innovation, but also their ability to provide a solution to the problem, or their complementary or supplementary nature.

It is possible that some ideas are not included in any of the concepts considered. Do not throw those ideas to the bin basket. They may be useful in future ideation sessions as idea triggers. In any case, the aim is to focus on the potential winners from a number of ideas.

Concept development example

Starting from the mind map in Figure 5.3 and focusing in some of the ideas collected in it (which ones?) we can develop the concept:

Dairy-based drink with additional ingredients to enhance its flavor, packed in a small brick container.

Concept development is an activity somewhat opposed to brainstorming. While in brainstorming we try to generate as many ideas as possible, without questioning them in terms of their possible real utility or assessing whether their real application is more or less feasible, in concept development an opposite approach is followed. We must combine the ideas in a product or service that we are going to test.

While the participation of people outside the development team may be possible and even desirable in brainstorming, only the development team should participate in concept development.

The way in which concept development is done can be quite similar to the way in which a jigsaw puzzle is solved. First, we can classify the pieces following different criteria. For example, in the case of the jigsaw puzzle according to the color of the pieces or the shape. A good strategy is to start by laying the edge pieces. Another good strategy is to group the pieces by color, such as putting together all the blue pieces of the sky, the green pieces of the field, etc. In the case of concept development, the way of proceeding is similar: we put together the different ideas taking into account relationships among them, looking for some criterion of affinity. Once we have made this classification (cf. Figure 5.4), we can start putting the pieces together looking for some reference element that gives us an anchor point on which to continue working. In the case of the jigsaw puzzle, it is usually a piece with a distinctive element that

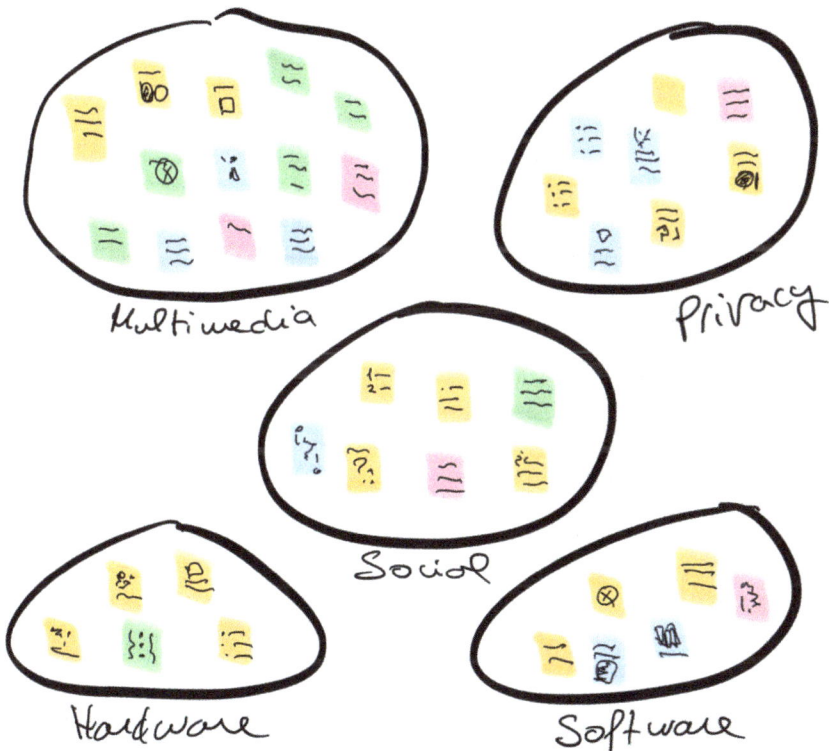

Figure 5.4   Concept development example: classification

allows us to place it in a certain position. In the case of concept development, it can be a clear and central idea. In both cases, this reference element is an anchor point on which to continue building, adding more tiles to the puzzle, and incorporating more ideas into the concept. In both cases, our construction will grow and consolidate until we have a result that is clear enough to be able to present it to test it and thereby validate it.

As suggested above, mind maps may also be created to drive concept development. Once an initial classification of ideas is completed, we can start putting the pieces together looking for some reference element that serves as an anchor point. In the case of the puzzle, it may be a piece with a distinctive element that allows us to place it in a certain position. In the case of concept development, it can be a clear and central idea. In both cases, this reference point serves as an anchor on which to continue development, adding more pieces to the puzzle, or incorporating more ideas into our concept. Our construction will grow and consolidate into a result appropriate to be introduced to stakeholders and be validated.

5.4.4   *The Now Wow! How? matrix and other classification methods*

Once we have one or several concepts developed, or even with the basic ideas, we can classify them according to a Now Wow! How? matrix (cf. Figure 5.5). This diagram has two axes, the vertical axis the degree of innovation, from a boring and already known end to a complete original idea, and the horizontal axis represents the difficulty of implementation, from a very easy end to an impossible end. This diagram provides an easy-to-follow method of evaluating the feasibility of ideas or concepts and their innovativeness. By arranging the ideas in relation to these two axes, we can visualize which are the most attractive ideas and those that are more feasible, while promoting group participation and consensus on them. Group members can use colored sticky notes, each representing one of the categories, on which to write ideas to arrange them on the diagram in turn.

Other kinds of diagrams can also be considered to classify concepts or ideas. The Four Categories Method enables us to classify ideas along a rationality axis, ranging from the most rational choice, to the most delightful, to the darling and finally to the "long shot." A best practice in this case is to decide upon one or two ideas for each of these categories. As a result, the team ensures to cover all grounds, from the most practical concepts to the most innovative solutions. Another method is the Bingo Selection method, where concepts have to be arranged in accordance to a variety of form factors, such as their potential applications in a physical prototype, a digital prototype, and a experience prototype.

5.4.5   *Idea selection*

Once we have done all the ideation process, it is time to select some of the concepts to continue with the next stage of the DT process: prototyping. A good way to proceed

Cool

Attractive ideas, but difficult to implement. Challenges

How?

Attractive, easy to implement ideas. The best ones!

Wow!

← Complicated → Easy

Unattractive, difficult to implement ideas. To be discarded.

These are not especially attractive, but they are easy to implement.

Now

Lame

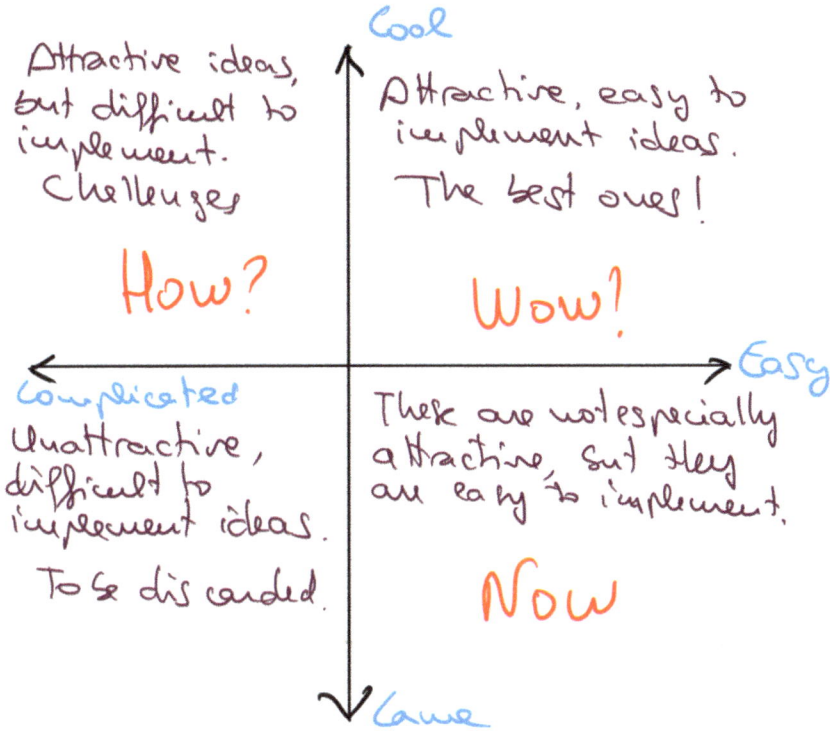

Figure 5.5   Now – Wow! – How? matrix

is following a democratic election of the wining concepts or ideas, following a post-it or dot voting.

First you need to write all the concepts on individual post-its. Then, each participant is given a number of votes, around 3 or 4, to vote for their personal favorites. Votes can be provided in the form of stickers or using a marker to make a dot on the post-it. This process allows every member to have an equal say. In case of a tie, a second round of voting can be held after a debate among representatives of the confronted ideas.

5.5   Conclusion

Ideation is a process to generate a large number of ideas and concepts to solve a problem, to satisfy a need or to face a challenge. For this, tools such as brainstorming, mind maps or concept development can be applied. The more ideas we obtain towards a solution, the better will be the chance to find the one that meets all our needs. To get a truly innovative idea, the ideas we generate have to be rich and imaginative, so initially we should not judge them. At this point, we look for ideas without questioning ourselves about the most suitable or brilliant one. The process should be inspiring enough to generate a large number of proposals from which we can select the best idea or a combination of several ideas. To finish this process, one or two candidate solutions will be selected and, based on them, concept or concepts that will serve as the basis for an innovative product or service will be developed.

Chapter 6
Prototyping
Manuel J. Fernández Iglesias[1]

A prototype can be something as simple as a sketch on a sheet of paper or a painted cardboard box. The goal of prototyping is to quickly visualize an idea and create a tangible model, which will improve communication, help to swiftly detect basic errors or misinterpretations of initial requirements, and improve the attitude toward the project of everyone who participate in it. The importance of this phase lies in the fact that we can have, in a short time and at a reduced cost, a solution proposal that can be tested by those people whose problems we intend to solve. It will always be better to confirm that the solution that we propose is correct when we are still in an early development phase, than to do it when we already have a final result in production. In other words, prototypes allow us to fail in a controlled environment, as many times as necessary and by limiting the costs of such failure, to finally obtain a satisfactory solution.

6.1 Introduction

Once we have completed the ideation stage [49] and we have at least one solution to the problem originally posed, in the form of a product or service, we are going to design and produce an object, the prototype, that allows us to visualize the final system and reason about it, together with the people who will eventually enjoy that solution (cf. Figure 6.1).

A prototype is a concrete representation of all or part of the expected outcome of our project, a tangible artifact and not a simple abstraction. It is an early, inexpensive and limited version that serves to reveal any undetected problems and facilitate our path towards a final solution. The prototype will allow us to evaluate the results as soon as possible, and thus detect any design problems or any misinterpretation of the initial needs or requirements.

Ultimately, prototypes allow us to make an idea come true, check whether or not it is viable to turn that idea into a product or service, and investigate how people think and feel about that product or service.

[1] atlanTTic, Universidade de Vigo [GID DESIRE], Spain

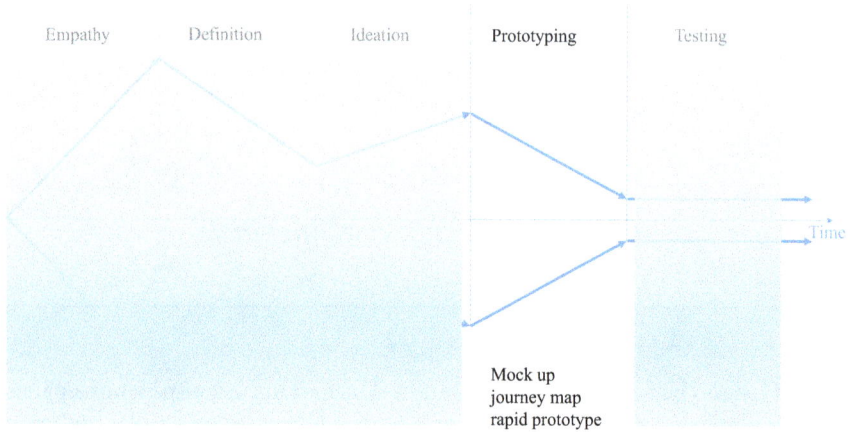

Figure 6.1 Prototyping stage in the DT process

6.2 The role of prototypes

From the perspective of design thinking (DT), prototypes are very important because they will serve as a common ground for talking with the people involved in the project about the solution we propose. The prototype is an object about which to talk, discuss, argue and make new proposals. To paraphrase the well-known adage, a prototype is worth a thousand images.

A prototype, especially when we can introduce it at an early stage of the project, also allows us to limit the cost of our mistakes. With a prototype, even when built as a preliminary representation of the final solution, a *low resolution* representation, we can quickly detect unwanted issues that we do not like or correct false assumptions or misinterpretations.

Prototypes may also lead to the development of new ideas in an incremental design process. They allow us to visualize the solution or final product at an early stage, and serve to build upon them what will be the next version, more advanced or more elaborate, of that solution.

Some keys to make good prototypes would be:

Start building something: The most effective way to start prototyping is by making a sketch. They are a fast and efficient way to organize concepts and ideas. If the prototype is a simplified version of the final solution, the sketch would be a simplified version of the prototype.

Do not get attached to the prototype: Prototypes are after all a preliminary representation of our project's target product or service. They are just something to help advance towards the ultimate goal. Prototypes are discarded once they have fulfilled their mission, so they cannot become a burden that conditions our subsequent work.

Identify the relevant elements: Be clear about what you want to show with the prototype. What actual questions do you have to answer? What specific part or functionality are we displaying with it?

Do not complicate yourself unnecessarily: The goal of a rapid prototype is not perfection, but to get something good enough to do its job at this point. It is not necessary to invest resources or time to go beyond that.

Do not lose sight of the user: We must always keep in mind the people for whom we are designing our solution. A prototype should always be tested with the target audience of the project in mind.

In the prototyping phase of an evolutionary and incremental design process, intermediate solutions are generated to address issues that will get us closer and closer to a final solution. In the early stages of a project, these questions can be more general, such as, "What impression did my proposal for an induction hob for a high-tech kitchen make?" At this stage, you have to produce inexpensive prototypes that are easy to make, but nevertheless will provide helpful and insightful feedback. In later stages, both the prototype and the questions can be a bit more elaborated. For example, in the final stages of our kitchen project, we can create a prototype that aims to discover whether users prefer to use voice commands or traditional control knobs.

A prototype can be anything with which users can interact, or at least that is capable of provoking reactions that indicate the perception that people have of our work: a panel with sticky notes, a physical object, a role game activity, a video clip simulating a commercial, a comic book storyboard, etc. Prototyping is about building something that allows users to somehow experience what they will meet when the project is complete.

If we think about a red vehicle, we can interpret these two words in multiple ways. *Vehicle* can refer to many different objects, such as a car, a motorcycle, an electric scooter, or even a spaceship. *Red* can indicate many different shades of a color, or even a political option for some audiences.

On the other hand, if we look at the image in Figure 6.2, although this sketch was drawn in a few seconds and can be interpreted in many different ways, it reveals an important number of specific aspects of this particular red vehicle such as, for example, that it is a car, that it has an antenna, that it is not a minivan, that it does not have a bicycle rack, that its lines are more those of a sedan than a sports car, etc. In turn, depending on how a specific person reacts to that sketch, we can obtain relevant information about that person, such as their preferences regarding vehicles, that there are people who carry their bikes on the car, etc.

6.3   Aspects of a prototype

We can study the prototypes from four different perspectives [50]: form, fidelity, interactivity and evolution.

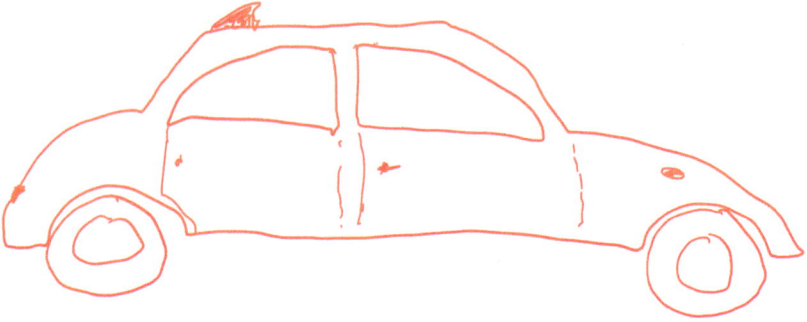

Figure 6.2   A red vehicle

6.3.1   Form

Depending on the nature of the solution that we are developing, such a solution can materialize in different ways. It can be a physical object made of metal, wood, plastic, etc. It can also be a computer program, a sketch, a video clip, a publication, etc. It can be very small, such as a prototype of a jewel printed in plastic with a 3D printer, or very large, such as a life-size model of an airplane.

On the other hand, prototypes can be materialized with the same formal elements as the solution that they represent, or with different formal elements. We speak of offline prototypes when their shape does not match that of the final product, and online prototypes when they participate in the same reality. For example, a paper sketch of an application using the *wireframing* [51] technique, a cardboard mockup or a computer-generated video of an object [52] are offline prototypes, and a simplified version of a program, a showcase, or a demo version of a product are online prototypes (Figures 6.3 and 6.4).

6.3.2   Fidelity

This aspect indicates the level of detail and functionality of a prototype. It reflects the balance between the relevant details that the prototype aims to show and the irrelevant details that the prototype leaves open to the imagination of the people who interact with it.

The fidelity of a prototype is usually related to its form. For example, sketches tend to be low-fidelity as they offer an overview with virtually no functionality, whereas demo versions or product pre-series are often high-fidelity, showing all or part of the final functionality in some detail.

In general, we talk about low-fidelity prototypes when they implement general aspects of the system without going into details, and about high-fidelity prototypes when they represent more specific aspects. High-fidelity prototypes serve, for example, to detail the entire interactive process of one or more specific tasks.

Figure 6.3 Model of the commercial area of the Vigo University campus made in just 1 h with office supplies (offline prototype)

For example, a life-size model of a car made of clay is a low-fidelity prototype that exposes general aspects such as the attractiveness of its lines, or its aerodynamic properties, while a prototype of a web page with the entire structure but with no actual content (e.g., populated with *Loren Ipsum* paragraphs) is a high-fidelity prototype when it comes to structure and a low-fidelity one with respect to the page content. Still, we can get an idea of what the final page will look like even without the prototype displaying the page's final content.

The advantages of high-fidelity prototypes are that they usually provide a lot of detail in terms of the functionality they show, they are usually interactive, and they can serve very well as a marketing research tool. However, they are expensive because they are often time-consuming to develop, difficult to modify, and may create false expectations. A high-fidelity prototype may be easier to understand for people who do not come from the world of design or have less technical knowledge, since they are visually closer to the final product.

On the other hand, low-fidelity prototypes are usually inexpensive, can be created quickly, and it is usually easier to interact with them. They are very useful for interface design and for identifying software requirements. However, they have significant limitations in terms of interactivity, and their perception is highly conditioned by the skill or style of the designer. They are usually offline prototypes. Low-fidelity

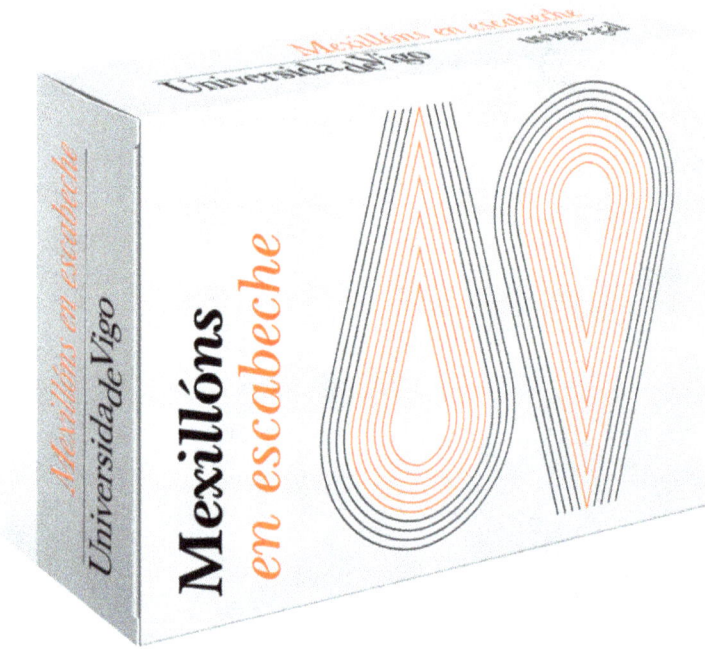

Figure 6.4   Prototype of a can of mussels. It is an offline prototype (i.e., a graphic design with no real existence) and it is low-fidelity (i.e., it only shows the external shape, although in high detail).

prototypes are appropriate, for example, for showing a concept design solution during the early stages of product development.

Depending on what we want to convey at a given stage and how advanced the development process is towards the final solution, a prototype may be a blend of low-fidelity and high-fidelity aspects. We can apply two strategies when building such a prototype. On the one hand, we can focus on a few characteristics of the final solution with a high level of detail (i.e., high fidelity), to evaluate a limited part of the final system in depth under real circumstances. In this case, we would be talking about vertical prototyping. On the other hand, we may identify all the basic features of the system, but without detailing the underlying functionality. In this case, we would be talking about horizontal prototyping. The latter is typically applied to the (high-fidelity) prototyping of user interfaces, whether of software programs or any other type of device or system (cf. Figure 6.5). For example, a prototype of a Web page can represent the structure of menus and navigation with a great degree of detail, but without allowing or supporting navigation through the different options, doing it in an erratic way or with broken links.

Figure 6.5   Horizontal prototype of a life-size wall thermostat. It is low-fidelity insofar its functionality is concerned, but the user interface has all the details. Do you find it comfortable to use? Does it seem intuitive to you? Are you missing something?

6.3.3   Interactivity

This aspect illustrates the ability of the prototype to support interaction. Based on this capability, prototypes can be classified into fixed prototypes, fixed-path prototypes, and open prototypes.

Fixed prototypes do not support any type of interactivity. They simply tell a story, they illustrate the operation of a system under certain previously established conditions. For example, a video presenting a new coffee maker or the animation of an engine's operation are fixed prototypes.

On the other hand, fixed-path prototypes support guided interactivity, limited to certain functions. For example, the prototype of a web page where only the registration process is active would be a fixed-path prototype. The only possible interaction will be that to carry out the sequence of steps that allows registering on that website.

Finally, open prototypes support interactivity in a way similar to the final system, but with the limitations of a prototype with a given fidelity. For example, a prototype

Figure 6.6   Software application prototypes: offline prototypes where interactivity is supported by moving sticky notes around when someone selects an option illustrated by the prototype

of a vending machine may faithfully display all the options available on the exterior control panel, but not include any of the interior mechanisms necessary to sell products.

According to this classification, we can see that a system is not more or less interactive because it is more or less dynamic. A fixed-path prototype or a video clip can be very dynamic, but not interactive.

6.3.4   Evolution

The evolution of a prototype refers to its path from its inception and across its development and use until it is finally discarded. Rapid prototypes are made in the early stages of a project, they are constructed relatively quickly and they utilize few resources. These are usually offline, low-fidelity, and are usually discarded after they fulfilled their purpose. Examples of rapid prototypes would be sketches, mockups, or wireline prototypes of computer programs, sticky notes prototypes or programs written in scripting languages.

On the other hand, iterative prototypes often require more resources and more time to complete. Normally its fidelity increases with each iteration. A special type of iterative prototypes would be evolutionary prototypes, which are characterized by evolving to become a part of the final system or the final system itself. As a consequence, they never are completely discarded.

6.4   Rapid prototyping techniques

The objective of the prototyping phase is to build a representation of a candidate solution that allows us to visualize and reason about it. In other words, we move from an idea to something concrete and real that, although it is not yet a final solution ready for delivery or commercialization, it does allow us to interact with aspects of the real system within the limitations of the prototype discussed in the previous paragraphs.

In fact, one of the key benefits of prototyping in project development is that it gives us the ability to test the results of our work early. Consequently, within the DT methodology, testing is an integral part of the development process, and the techniques used in the testing stage are not something to apply only at the end of the process, but are part of the development process itself.

In the following paragraphs, we collect three of the most popular and effective rapid prototyping techniques. We begin by discussing sketches, already introduced above, and then describe models as an evolution of the former. Finally we describe the *Wizard of Oz* technique, as an approach that encompasses all the techniques described.

6.4.1 Sketches

Especially in the early stages of development, in most cases, the best approach to prototyping is a low-quality, low-fidelity prototype, a prototype that is made quickly and with few resources to show a general idea, a sketch. No complicated technological devices or state-of-the-art materials are necessary. A pencil and some paper, a piece of clay or expanded polystyrene foam (EPS) and a blade is more than enough.

In case that we are developing a computer application or a protocol or service, it is also possible to apply a similar strategy. For example, we can use sheets of paper and sticky notes to prototype navigation through a computer application, wire-framing techniques, or simple diagrams and graphics to illustrate the deployment of an application or service (cf. Figure 6.7). It is easier for a person to focus on images than on text descriptions. In a typical software development case, a design team may produce a series of paper prototypes that can be gradually worked on to demonstrate how certain tasks or problems are addressed. In the case of developing tangible devices, such as a computer mouse, the design team may use a number of different materials and ergonomic shapes to allow them to test the underlying technology.

6.4.2 Mockups

Mockups or models are another very popular rapid prototyping technique, as they address the aesthetics of the final system, its ergonomics, and even help to identify possible production problems of the final product. Mockups are an instrument to explore the form, composition, and functionality from an idea to detailed design.

The systematic introduction of models or mockups as a prototyping technique started in 1947, when Chuck Yeager was designing the Bell X-1 supersonic aircraft. For this, he used 50 in.-caliber projectiles to emulate that aircraft and thus studied supersonic flight. Interestingly, the Bell X-1 became known as *the bullet with wings*. After this historic event, designers and engineers introduced model making as standard practice to bring their sketches to life in the three-dimensional world.

Although computer design has become widespread and 3D rendering and virtual reality technologies have evolved dramatically, physical models are still essential

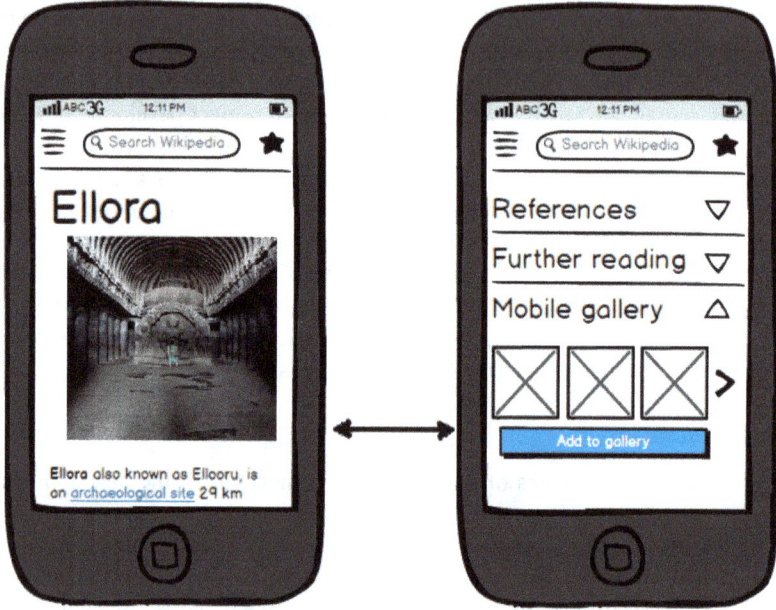

Figure 6.7   Mockup prototype of a mobile application of a virtual art gallery (Maryana Pinchuk y Jon Robson)

in many fields, especially when physical interaction is required to fully understand whether a product is acceptable or not.

An example that perfectly illustrates the importance of a good prototype on time is the introduction of the new official NBA ball in June 2006 [53]. The new ball featured a new design and a new material that, according to the manufacturer, offered better grip, feel, and consistency than the official leather ball of previous seasons. It was the first major change in more than 35 years, and the second ball in 60 seasons. In October 2006, the new ball began to be used in official matches. It turned out that the ball caused injuries, and players ended up denouncing the NBA for safety problems at work.

In December 2006, the leather ball from the previous season was reinstated. Most probably, the lack of a mockup or model featuring the new material for players to experiment with it, introduced at the early stages of the design process, was instrumental to the failed introduction of the new basketball.

6.4.3   Video prototypes

Video prototypes are especially suitable when we want to show the characteristics of a system with great realism at a still very preliminary stage in the development of that system [54]. With a good video clip, it is possible to show how people would interact with some physical device even when that device does not exist yet. Through this kind of prototype, it would even be possible to demonstrate something impossible with current technology, or located in an unattainable place (e.g., on another planet, in an inhospitable place, in an artificial environment, etc.), and therefore study people's reactions to it. With this, we can study the reactions of the people interested in our project or of the team members in a very preliminary stage of development, to discover possible usability problems, errors or misinterpretation of the needs of final users.

Obviously, the possibilities of interacting with a video clip are very limited. Video prototypes would be fixed path prototypes, since with them it is possible to show or illustrate how a system interacts, but only for the interaction sequences filmed on video. In addition, the level of detail of the systems represented in a video is usually limited, as the most superficial or visible aspects of that system are the ones represented.

To make a video prototype you do not need too many resources. Most likely, all or almost all team members will have a smartphone with at least one camera capable of recording video in high definition, with basic video editing functionality, and with the possibility to transfer these videos to a laptop or desktop computer, as well as to share them via the cloud or social networks.

On the other hand, we must always keep in mind that we are making a proto-type, and therefore it is something ephemeral bound to be discarded as our project progresses. The amount of resources that we dedicate to video prototype production must be consistent with the role that our video prototype will play.

Making a digital video prototype is a process similar to any video production project [55], which can be organized into three main phases, namely preproduction, production and postproduction.

Preproduction consists of planning our video creation project and getting all the information and resources that we are going to need to make the video prototype a reality. The planning of our video prototype is probably the most important task of all, since the product of this phase will determine the final result in a decisive way. Among the key aspects to take into account at this stage, we have to define the duration of the video prototype, at least approximate. It has to be long enough to convey everything we need to convey, and short enough to hold the attention of our audience. Advertising videos are usually less than 30 s long and corporate videos rarely exceed three minutes. If we are able to create interesting, inspiring or provocative content, our audience will tend to stay tuned for longer.

Related to duration, we must also estimate the number of takes necessary to complete the video. We also have to identify all the necessary resources beyond video takes at the next phase, such stock video clips, as animations, diagrams, and the elements in the soundtrack, that is, what music or sound effects will be necessary based on the content and our audience.

It is also very important to write a script or storyboard where it is described in detail what we want to convey to our audience. In this sense, our video prototype can be an evolution of a previous prototype in the form of a sketch or a storyboard.

Production is the process of obtaining the image and sound of our video prototype, guaranteeing that all the necessary elements for it are available when required. First of all, we must familiarize ourselves as closely as possible with the technical aspects of our audio and video recording equipment. In the case of making a prototype with a mobile phone, the controls for recording and storing videos will be relatively simple, but even so, we must familiarize ourselves with the different supported resolutions, the storage formats for both audio and video, the relationships of supported aspect ratio, methods for sharing and distributing shots, whether there is any way to control lighting, focus, and framing (e.g., whether it is possible to zoom), etc.

In general, our takes should be correctly presented, well focused, and well framed. We must bear in mind that framing is our means of expression. Everything that is out of the picture frame will not exist for our audience. We must also bear in mind the sources of light and the shadows cast by objects and people. In addition, takes should be as stable as possible trying to minimize or completely avoid vibrations or unwanted camera movements. In terms of sound, the microphone should be as close to the subject or sound source as possible, while avoiding unwanted noise or sound sources. In the case of using the mobile camera, we must assess the convenience of taking sound shots with an additional recorder or with other cellular phones.

Finally, we must have the ability to identify the potential of a good shot, although that is something that is acquired with practice. An interesting strategy would be to film more takes than we might initially think we need, or take longer takes to have more flexibility in transitions during post-production.

Finally, postproduction corresponds to editing the video from the resources we have obtained in the production phase, as well as coding and packaging the different elements of the finished video in the appropriate container (image, soundtracks, subtitles, metadata, etc.). This is the phase when we will really tell our story. Our audience must understand and assimilate what we want to convey about how we are going to solve a certain problem. Therefore, clarity takes precedence over artistic experiments. Using all the material obtained in the production phase, we select everything that is relevant to illustrate the operation of the system or service that we are prototyping, and we arrange it according to a temporal order that effectively transmits that.

Editing takes as a reference the script or storyboard created before shooting the video. With that initial outline, the takes that will be part of the final prototype are selected and edited in case it is necessary to adjust their duration, framing, etc. We will eliminate anything not relevant, focusing on those takes or segments that will best reflect what we want to convey. In the same way, we will select the appropriate audio cuts (i.e., music, voice-overs), additional resources (i.e., stock material, animations or diagrams) and we will arrange them in the corresponding place in the final video prototype. It is also the time to integrate other elements such as titles, special effects, and transitions.

The requirements or limitations related to the dissemination of the video are also identified here, as well as the tools necessary to convert or transcode the material to its final format. For example, we will have to take into account whether the video prototype will be broadcast in streaming over the Internet, whether it will be distributed on a physical medium, whether it will be displayed on a large screen or on a mobile device, etc.

It could be argued that conveniently completing the three previous stages will most likely provide a good video prototype. However, it takes good training and some experience to master the various phases of making a video. In fact, there are many training programs and professions that focus on very specific aspects of the process (cameras, sound engineers, editors, producers, filmmakers, scriptwriters, illuminators, etc.). However, with some basic notions, covering the fundamental aspects of the digital video creation process would be enough to face the challenge of producing a video prototype suitable for the prototyping stage of a DT project.

6.4.4 Wizard of Oz prototypes

Finally, there is a technique that combines all the rapid prototyping techniques described in the previous paragraphs named after the 1939 movie *The Wizard of Oz*. This name is attributed to the usability expert Dr Jeff Kelley, who was inspired by the scene in that film in which the dog Toto opens a curtain to discover that the Wizard is not actually such a wizard, but a man who is operating a set of buttons and levers to bring to life a representation of a magical character. This technique, called also WoZ after its acronym, is widely used for low-fidelity prototyping of systems that interact with human users, especially natural-language interaction systems [56] [57].

WoZ prototyping is based on a sketch or a rudimentary mockup of the final solution. The prototype can be quite simple and be based on every day or widely available objects to represent specific parts or functions of the final solution. After the prototype has been created, role-play is used to illustrate how the prototype is interacted with. In addition to the sketch or the mockup, a WoZ prototype requires a script with the instructions for the *wizard*, that is, the person who will perform the tasks that will simulate the behavior of the final product. The person interacting with the prototype may or may not be aware that the assistant's tasks are performed manually by a human being rather than by a machine or computer application.

For example, to develop a WoZ prototype of a paddle court telephone booking service that allows clients of a sports center to make or cancel a reservation, we would first build a tree with all possible requests of the caller and the responses of the system, trying to take into account all possible scenarios (cf. Figure 6.8).

Once the questions and answers tree is built, that is, the actual WoZ prototype, we would invite the target users of that service to test it. To do this, the person who acts as the wizard would start the conversation by simulating the service's welcome announcement, and would react to keypad interactions from users according to the script.

We can see that there are several elements to be considered when developing the prototype, such as the level of exposure of the wizard (i.e., what does the person who interacts with the booking service know about the hidden assistant? Does they know their existence?), or its role (i.e., what part of an application does the hidden wizard emulate?). We can also utilize this prototype to study aspects such as the acceptability of the assistant, to streamline the handling of most common situations, or to address issues not identified during the system's development.

6.5 Team challenges

We discuss below a collection of prototyping activities designed as competitions among teams of people. Although these are essentially rapid prototyping exercises,

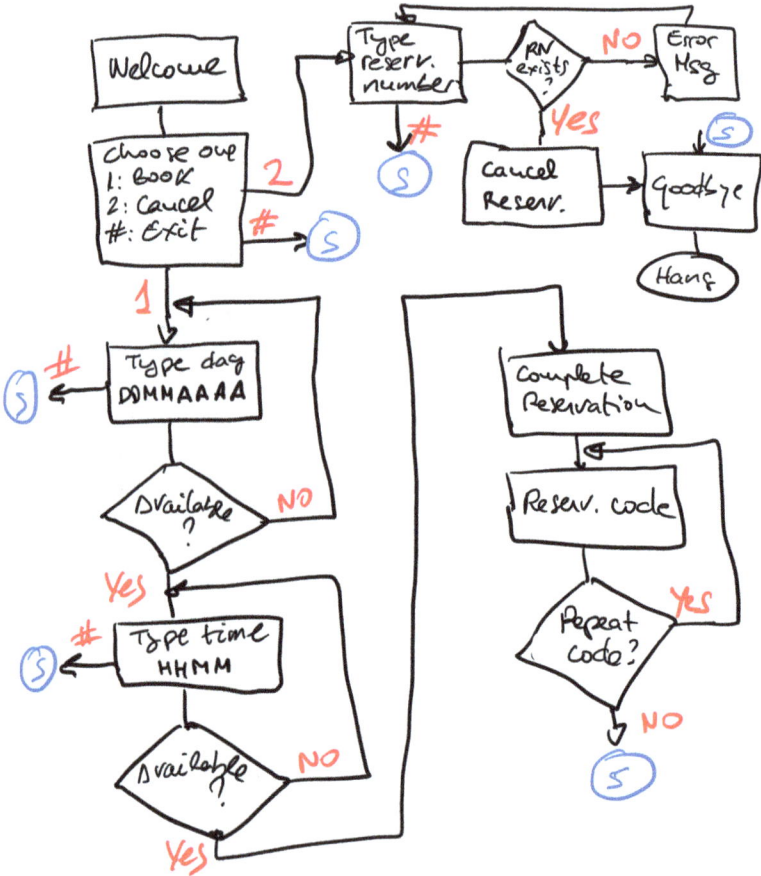

Figure 6.8   *WoZ prototype of a telephone booking service. The system produces vocal cues, and users interact through the telephone's numeric keypad.*

they are really appropriate for other phases of the creative process, such as, for example, as instruments to promote creativity and the generation of innovative ideas. They can also serve as interesting exercises in team building processes, for example to identify aspects of the personality relevant to teamwork that serves as a basis for assigning roles.

In addition, these exercises demonstrate that even the simplest tasks may have multiple solutions. They help to convey the idea that making mistakes is not a bad thing, and that many times you have to experiment with multiple alternatives until you reach a final result.

These challenges are based on the manual construction of a structure with the materials provided, based on the ideas developed within each group. The *challenge* nature implies that participants are facing a competition, and that therefore there must be a reward for the winning group, even if it is a symbolic one (e.g., a diploma, a bag of candy, etc.). The reward also serves to express that you are aware of participants' commitment to the task at hand.

Each challenge most be adjusted to the characteristics of participants and the selected venue. All proposals can be carried out outdoors, but sometimes an indoor alternative may be more convenient due to unexpected weather conditions, such as rain or gusts of wind.

6.5.1 The paper tower

The challenge is to make the tallest tower possible using only newspaper or similar. Participants are provided with a certain amount of backward press or wrapping paper, and are instructed to design the tallest tower possible using that paper as the only material, and build it. The use of any other construction material or joining elements (glue, tape, clips, etc.) is not allowed.

The time available to complete the challenge can vary depending on the specific context in which the challenge takes place. A reference duration could be 30 min.

The team that makes the tallest tower wins. In the event of a tie, the team that uses the least paper wins.

Generous, but not unlimited, amounts of paper are important for this challenge. The idea is that participants do not feel conditioned by the availability of materials to develop their ideas, but do not try solutions based on brute force, such as towers built simply by stacking paper.

6.5.2 The flying egg

The challenge consists of dropping an egg from about 3 or 4 m and get it to land without breaking. For this, the teams must build a means of transport for the egg that guarantees a descent without incidents.

The team that manages to land the egg safely wins. In the event of a tie, the team that makes the trip slower wins, that is, the team that makes the egg stay in the air the longest. In the event that there are no surviving eggs, additional time can be provided for teams to improve their means of transportation.

To build their vessels, each team has to choose four items from a bill of materials, such as:

- Two rubber balloons.
- Two plastic cups.
- Five rubber bands.
- A block of modeling clay.
- Five toothpicks.
- Five wooden paddles.
- Five erasers.
- Eight clips.
- One meter of rope.

In the example list above, each item counts as one. Thus, a possible combination would be a kit consisting of two balloons, two plastic cups, five rubber bands, and a meter of rope.

Also, a generous amount of chicken eggs are needed. You may limit the number to two or three per team, or give out unlimited eggs. Alternatively, the challenge could be performed with only one egg per team, limiting the total time of the challenge. In this case, if there are no surviving eggs, the team that manages to keep the egg in the air the longest wins.

You also need a stopwatch to measure time (e.g., the timer of a smartphone) and a blackboard to record times and failed attempts.

Minimum rules must be established to avoid distorting the challenge such as, for example, that at least 50% of the egg should be clearly visible, and thus avoid solutions that involve wrapping the egg in shock absorbent material.

This challenge should be carried out abroad. In addition to issues related to cleanliness – this exercise tends to be a bit messy due to debris generated by unsuccessful attempts – other factors such as wind, orography or soil characteristics can make it more challenging. In addition, it is easier to find places with enough height to complete the drops.

6.5.3 *The tennis ball*

This challenge consists of building a structure to guide a tennis ball from a high point, such as a table, to a box or bucket placed on the ground. The challenge is completed by achieving an additional objective, such as putting the ball in the bucket as many times as possible in a given time, or achieving a given number of repetitions as fast as possible.

The time provided to build the structure can be around 20 min, and the additional time required to complete the challenge will depend on the specific challenge that we define. For example, if we set a maximum time of 1 min and the competition is based on achieving the highest number of repetitions in that time, the total maximum time will depend only on the number of participating teams.

As in the balloon challenge, there are no strict rules regarding the materials needed to accomplish the task. Among the appropriate materials for this challenge we may propose:

- Adhesive tape.
- Corrugated cardboard.
- Wooden or plastic sticks.
- Aluminum foil.
- Cardboard boxes.
- Cardboard or plastic tubes.
- Drinking straws.
- Rubber bands.

There are several ways to distribute the materials: each team receiving the same kit; that each team may choose a limited number of items from the list as in the previous challenge; each team may use each item exclusively for a limited time, and then return the surplus to a common deposit; etc. This last option promotes more diverse constructions, but you must be prepared to mitigate conflicts that may appear among teams. The absence of adhesive tape or some other bonding element greatly complicates this challenge.

Before the challenge, the bucket is fixed to the ground at the same distance from the table for all groups, and it is not allowed to change the position of the bucket or table. In addition, as in the previous case, rules that help not to distort the challenge should be defined, such as:

- Initially, the ball is placed on the table at a fixed point marked with an X. A member of the team must push the ball from that point.
- No member of the team may touch the ball from the launching pod until it falls into the bucket. The ball must be in motion by itself from the starting point to its final destination.
- Moving does not just mean rolling; the ball is also allowed to position itself by its own inertia or by means of any other object, standing or moving.
- The ball must be brought to the spot marked with an X on the table before the next attempt. It is only allowed to touch the ball during its transfer from the bucket to the table, or from the ground in case of unsuccessful attempts.

The most critical element is to get the ball moving. The above rules ensure the construction of some kind of structure and avoid converting the challenge into a basketball game.

This type of challenges requires not only building a prototype but also testing it in a competitive environment. Some teams focus on construction, while others focus their efforts on the actual competition. The establishment of roles among the team members is also common, such as a person appointed to take the ball from the bucket, pass it to a second team member, and a third one launches it from the table. That way, precious time can be saved on each try.

6.5.4 The candle trip

This proposal takes time, and also requires a lot of resources, but usually it turns out to become a memorable experience. Also, it illustrates the prototyping and testing process very well.

The aim is to build a water transport for a lit candle and make it navigate a channel in the shortest possible time without sinking. Obviously, one or more waterways are required, generally one for every four or five teams. Each channel should be about 3 m long and about 30 cm wide. A depth of 2 or 3 cm would be enough. Finally, tabletop fan at one end of the channel is also needed.

As in the flying egg challenge, we will need a stopwatch and a board and marker to keep track of the scores.

In this case, the kit of materials may include items such as:

- Sheets of paper.
- Corrugated cardboard sheets.

- Adhesive tape.
- Scissors.
- A cutter or knife.
- A bottle of white glue or universal glue, or a hot glue gun/pen.
- A lighter or a box of matches.
- Paper clips.
- A roll of aluminum foil.
- Wooden sticks.
- Drinking straws.
- Bottle corks.
- Small paraffin candles.

To distribute materials among teams, we can follow one of the strategies indicated in the tennis ball challenge.

Each group will have about 25–30 min to build their water transports. The vessels must be capable of carrying a lit candle from one end of the channel to the other. Teams can make as many attempts as they want, or a limit for the number of attempts may be defined. Before each new attempt, minor adjustments are allowed within a limited time (e.g., 1–3 min).

Before starting the challenge, the fan is turned on. A try begins by placing the vessel with a lit candle on it at the end of the channel where the fan is located. It is not allowed to touch the candle transport after placing it on the water, and the candle must remain lit when the boat reaches the other end of the channel.

Different criteria may be used to assign scores. For example, the winner can be the group that completes the challenge in the shortest amount of time, regardless of the number of attempts. We may also take into account the number of attempts or the total navigation time from the start to the finish line.

This challenge can also be carried out using land vehicles, which eliminates the need for water channels. In this case, a straight race track of about 3 m can be constructed using paint or adhesive tape. Besides, the bill of materials may include items that can be used to build wheels or elements that enable vehicles to slide on the ground.

6.6   Conclusion

Prototyping in DT is the iterative generation of artifacts intended to answer questions that get you closer to your final solution. In the early stages of a project those question may be broad – such as "do my users enjoy cooking in a competitive manner?" In these early stages, you should create low-resolution prototypes that are quick and cheap to make (i.e., in the range of minutes and cents respectively) but can elicit useful feedback from final users, team colleagues and other stakeholders (e.g., funders, regulators, retailers, etc.). In later project stages, both your prototype and questions posed may get more and more refined. For example, a later-stage prototype for the cooking project may be constructed that aims to find out whether users enjoy cooking with voice commands or visual commands.

Prototyping and testing are tasks that should be carried out concurrently more than two different activities between which the designer transitions. What you will be trying to test eventually and how you are going to do it are critically important aspects to consider before creating a prototype.

However, addressing these two activities together also brings in new perspectives. Though prototyping and testing are sometimes entirely intertwined, it is often the case that planning and executing a successful testing scenario is a considerable additional step after creating a prototype. Do not assume that you can simply put a prototype in front of a user to effectively test it. In many cases, the most informative results will be the consequence of careful thinking about how to test in a way that users are motivated to produce the most natural and honest feedback.

Chapter 7
Testing

Enrique Costa-Montenegro[1] and Francisco J. Díaz-Otero[1]

In this chapter, we will talk about the testing phase, which is the final phase of a design thinking (DT) process, as illustrated in Figure 7.1. In it, we will use prototypes to study how the people for whom we develop the solution interact with them. This way, we can reveal new solutions to existing problems or find out if the solutions we develop were really successful or not. The results generated from these tests are used to redefine one or more of the problems established in the previous phases of the process, and to understand in a clear and complete way any unforeseen situation that could appear when interacting with a product or service in a real environment.

7.1 Introduction

When you want to determine and understand exactly how people will interact with a product or service, the most obvious method is to test it in a real environment. In any case, it would be very cumbersome to wait until you have produced something completely finished for people to try it out. Take the example of the new NBA balls that we presented in the Prototyping chapter (cf. Section 6.4.2). As we saw then, the most reasonable strategy is to develop simple and reduced versions that can then be used to observe, record, judge, and measure the specific elements of the different components of a solution, as well as people's interactions and reactions. The latter is basically the testing phase.

What we are trying to test, as well as how we are going to test it, there are critically important aspects before starting the testing phase. Although prototyping and testing should be interrelated, on many occasions, the careful planning of a test scenario is a preliminary step that should be taken into account even during the creation of a prototype. It should not be assumed that a prototype can simply be placed in front of a person for testing. In our opinion, the most informative results will come from careful thought about how to conduct the test in a way that will elicit the most natural and honest reactions and feedback. Moreover, it may sometimes be necessary to make intermediate prototypes during some of the previous stages of the DT process.

[1]atlanTTic, Universidade de Vigo [GID DESIRE], Spain

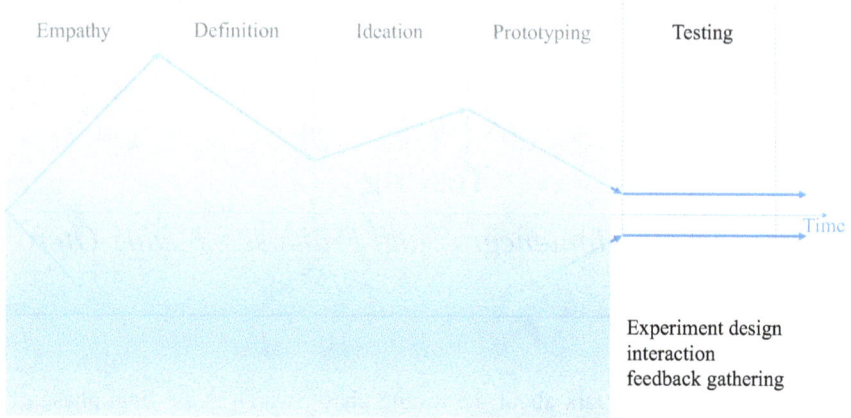

Empathy   Definition   Ideation   Prototyping   Testing

Time

Experiment design
interaction
feedback gathering

Figure 7.1   Testing phase in the DT process

In short, once the prototype has been made, we proceed to rigorously test it, but not before carefully planning this test.

This is the final stage of the DT methodology discussed in this book. In any case, in the iterative process proposed by this methodology, the results generated during the testing phase are used to go back and redefine the problem, to try to better understand the context in which we develop the solution, how people think, or how they behave.

We can say that we test the prototype with three complementary objectives:

1.   On the one hand, we seek to refine the prototype itself to get even closer to the solution that the people for whom we develop our project really expect. In other words, tests inform and guide the next iterations of the project. Sometimes this means going back to the design board and retaking any of the previous phases of the DT methodology. Testing can also help to identify previously unidentified problems.
2.   In addition, we use testing to learn more about people. Tests are another opportunity to generate empathy through observation and engagement, as unexpected ideas often appear. If people experience difficulties with the prototype they are testing, the team should review its list of possible solutions and strategies to establish new ways to solve the same problems.
3.   Finally, tests serve to refine the point of view created during the definition phase. In some occasions, tests may reveal that the selected challenge, the problem to be addressed, was not correctly identified.

Feedback obtained during testing is very valuable. If we do not have a good understanding of people's needs to carry out their activities and tasks, the iterative design process and solution will fail. Just as with each stage in a DT process, the tests should provide new insights to improve understanding and to help the designers to define or redefine the various problems we face. Therefore, it is necessary to obtain feedback whenever possible, to test with the people to whom the solution is addressed, and to analyze the results to determine that it is working well it and it will not eventually cause problems.

As we pointed out before, testing sessions are more fruitful when they are carefully planned and organized. For this, we can make use of the tools described in the following sections. In addition, we can use the team challenges introduced in the Prototyping chapter (cf. Section 6.2) to demonstrate the iterative process by which we refine a prototype using feedback from the testing phase.

7.1.1 Why testing?

The testing phase is the opportunity to present a product to the world, to test it in real life and real time. During this phase, the developers have the opportunity to see if they have addressed the problem in the proper way. In this sense, three different aspects are covered while testing:

- The team can generate feedback from users, specific to the prototype, and this feedback in turn deepens the knowledge about the users.
- The team will find new ideas to feed into all stages of the process during iterations.
- Finally, observation during this stage is likely to uncover needs that users have never identified before.

In this phase, tests are carried out with the prototypes previously made and users are asked for their opinions and comments based on the use of the prototypes. This is an essential phase in DT as it helps to identify errors and possible shortcomings of the product. Based on the tests, various improvements to the product can be presented.

Despite being the last phase, the team may encounter several situations that require going back more than one stage in the process. For example, if the team realizes that the problem is not well defined, then it is necessary to go back to Definition and start again from there. Otherwise, the team will most likely go back to the prototyping phase to refine certain details or include new features.

Testing is, therefore, the last phase of a DT process. It is the "moment of truth" in which we will show the user what we have designed for him. It is the end of a process of generating ideas, which have been landed on in the form of a prototype. Ideas that have been based on a previous research and definition included aspects of special value for the user.

7.2 How to address the testing phase

There is a wide range of testing methods available during a DT project. At the heart of all methods is the need to test the solutions made real through prototyping. The best is to use a natural environment, that is, the normal environment in which people would use the prototype. If testing in a natural environment is not possible, people will not act in a natural way. In this case, we should try to guide the people who are going to perform the test to carry out specific tasks or to act out a specific role when testing the prototype. The key is to get these people to use the prototype as they would in real life, as much as possible.

Several aspects have to be covered in the testing phase. Once fulfilled, the team should plan the test, assign roles to each member in the context of the testing and properly define how to receive the user feedback from the prototype.

The main aspects to be covered are:

1. **Context of the test**: Ideally, the prototype should be tested in a real-life context. For example, in the case of a physical object, we could ask people to take it with them to use it within their normal routines. For a user experience, we would try to create a scenario in a location that captures the real situation. If it is not possible to test a prototype on-site, we would look for a more realistic situation by having users take on a function or task when they approach your prototype.
   Performing a test is not as simple as taking the testing subjects and the prototype into the same room and observing what happens. To obtain the most useful information from a test, there are several aspects that we must take into account.
2. **Focus on the prototype**: The prototype is to be tested, not the people. The prototype should be designed with a central question in mind, a question that will be tested at this stage.
3. **Plan the context and scenario**: As far as possible, the scenario in which the final product or service is most likely to be used should be recreated. Thus, more information can be obtained about the interactions between the people, the prototype and the environment, and about the problems or new situations that could arise as a result of these interactions.
4. **Correctly inform test subjects**: We have to make sure that the people who are going to participate in the test know what the prototype and the test are about, and understand what is expected of them. If it is impossible to carry out the test in a real environment, we must clearly explain the role that each person participating in the test must play.
5. **Observe and capture feedback**: We have to make sure that we do not interfere with the testing process while recompiling comments or reactions. We must recompile these comments and reactions in such a way that we can observe what is happening without interference.

If we are careful with these five aspects, we will be able to extract the maximum possible information from the testing process. Next, we will discuss in more detail the planning process of a test, and then provide some techniques to optimize the capture of information from the people who participate in the test, that is, to obtain feedback about the test.

7.2.1 Test planning

As explained previously, testing is not simply showing the user our prototype. It implies carefully listening, detached from our own ideas and prejudices, to the user's output. The end of this phase also marks a turning point and a strategic decision point. We will have to decide, according to the feedback collected from the user, what actions we are going to take in order to continue to get closer to the solution that fits their needs and desires.

The moment of testing brings us back together with the user, just as it happens at the empathy phase. It is important to be prepared for this moment. One way to do this is to review with the team how the process evolved up to this point, taking some notes in relation to each of the previous phases:

- Empathize: Select some of the phrases that the users told us or a feature of their behavior, and see how they relate to the solution we are going to present to them. For example, let us imagine that we have designed an App that allows them to control their expenses, of which we are going to show them some screenshots. Before the interview, we will write down some of the phrases that they told us and that we think have led us to the design of the functionalities that we are going to present to them.
- Define: Take note of the focus of action that led to each of the ideas landed on in the prototype. Continuing with the previous example, let us imagine that one of the action points we defined was "I find it hard to save without motivation." Among the functionalities we included in the App based on our ideas, there is one that allows them to choose a "purchase goal." As they save, a bar fills in the money they need to reach that goal. This solution connects with the aforementioned action focus. Having all the Action Focuses-Generated Ideas relationships in front of us in the validation will help us to better analyze the user's feedback in relation to our work.
- Ideate: In this case, we will write down the ideas. And how they have been materialized in the prototype. During validation, we will pay attention to user feedback. Observing if the power of the idea has been increased or reduced. From the way it has been landed or intends to be landed.

Some other important aspects to take into account are those related to organization and tools to keep track and record of the interviews and feedback received from the users. For instance, some useful tips and tricks are:

- Presentations are fundamental tools in the context of testing. They are useful to correctly inform the test subjects of what is expected from them, to present the prototype, or even to perform the function of a prototype. They are also useful to collect and disseminate feedback obtained during the testing process.
- Keeping notes and having a simplified outline of our route can help us to get feedback. In an orderly way, we will have in front of us the logic of the steps we have been taking. And we will be able to compare it with everything the user tells us.
- When talking to them, we must never forget our main objective: to obtain the most genuine information possible to know if we are connecting with their needs and desires. To do this, one fundamental motto is *to listen, to understand rather than to sell*.
- For this reason, when we show our prototype, we must listen. Put aside our desires and expectations. And remember that the user is at the center of the whole process.

- Selling is on the opposite side of listening and understanding. And we should not do it. Our efforts should be focused on perceiving the coherence between the user's words and his/her actions. In detecting the brakes that arise when interacting with the prototype. And in answering the questions, always looking for them to lead us to more information provided by the user.
- As a last point, the ego is the main enemy of listening. It is responsible for blaming the user when he/she does not respond as we would like. And the one that can lead us to abandon the process. Even in cases where the user is not clear, or contradicts himself/herself, we must put ego aside. And strive to understand the reasons why we are not managing to empathize or connect with the user.
- We should prepare the testing environment so that we can focus on showing the prototype with a neutral attitude about it. As designers, we may be tempted to give our opinion about the benefits of our creation, but we must let people experience the prototype and draw their own conclusions. Avoid over-explaining how the prototype works or how it is supposed to solve a given problem. It is better to let the experience of using the prototype stand for itself. We should limit ourselves to observe without interfering.
- On some occasions, it will be convenient to create multiple prototypes, each one with a change in some of the aspects we intend to test, so that the test subjects can compare the prototypes and decide which ones they prefer and which ones they do not. In most cases, it is easier to explain the likes and dislikes of the prototypes when you can compare them.
- We have to ask the people participating in the test to express what goes through their minds when they are exploring and using the prototype, to speak through their experience. It may take some time to get them relaxed and adjusted to an unusual situation, so it may be a good idea to talk about an unrelated topic and then ask them questions such as *What are you thinking about right now while you are doing this?*
- We must always bear in mind that we are fundamentally observers. We pay attention to how the prototype is used, regardless of whether its use may seem correct or incorrect. We will try to resist the temptation to correct if we misunderstand how the prototype is supposed to be used. Mistakes are valuable learning opportunities.
- In this sense, the questions we ask during the tests are very important, even if you think that we already know what a specific user will answer without the need to ask. We should ask questions such as *What do you want to say when you say …?* or *How did you feel about that?* and, most importantly, do not get tired of asking *Why?*

7.2.2   Assign roles

In order to prove our prototypes in front of an audience, there are several ways, after the careful planning realized in the previous section, to assign roles in the group to conduct the test. Generally speaking, user testing involves selecting users from a segment of the population and achieving insights about our prototype. However, depending on the agenda, deadlines or budget of the project, testing the ideas frequently with as many people as possible is not always feasible. Therefore, it is also useful to do some prototype testing with people you know, making an effective technique for validating ideas with minimal efforts. This technique of running a quick test with 5–6 people who the team knows and observing how they use, manipulate and deal with the prototype, is a fast, simple and cheap way to learn a lot of insights.

Another important point to take into account is the type of users recruited for the testing (age, media channels, social channels, etc.) and if you are planning an in-person test or remote test. Selecting a representative group of the population that will be the final users of your prototype is essential to perform an objective testing. Getting user feedback following carefully the test planning explained in the previous section is the core of this stage of DT.

In any case, not only final users of our prototype will play a role in testing, we must select others as important as those defined. These basic roles are:

- Host or moderator: Provides a short and concise introduction to the context and scenarios. Without much explanation, the user must also discover by himself/herself all the aspects of the prototype. He/she also guides the questions.
- Players: They accompany the user to create the prototype experience.
- Observers: They just watch and observe the user experience with the prototype and annotate feedback.

7.2.3 How to get user feedback

The testing procedure should be strictly followed by the team members, each of them according to the assigned role:

1. Let the user experiment with the prototype. Show it, do not explain it. Provide just a minimum amount of information so they can understand it. Do not clarify what it does or provide any rationale about the functioning of the prototype.
2. Let the user vocalize while living the experience. Ask *"Tell me what you are thinking while you are doing this"* style questions.
3. Actively observe how they use (or misuse) the prototype. Do not correct what the person is doing, just observe.
4. Always ask questions. This is the most important thing in this phase.

Compiling information and comments from people is a characteristic aspect of the DT methodology that is of crucial importance in the testing phase. To maximize the benefits of information capture, it must be carefully planned to avoid making mistakes. We can identify several aspects to be taken into account during the capture of the comments and reactions of the participants in the tests.

How to request comments: It depends on the type of prototype that was built. As we pointed out before, it may be convenient to test several versions of the prototype. This facilitates the appearance of critical comments because people tend to abstain from openly criticizing the prototypes if only one version is available. When users can choose between alternative versions and allow them to compare the different prototypes and say what they liked and disliked about each version, they will get feedback that is more honest.

Choosing the right people: This is necessary to ensure the usefulness and relevance of your comments. If you are in the early stages of a project and only want simple and approximate feedback, testing the prototypes with team members would be sufficient. Towards the end of the project, when the prototypes become more detailed and closer to a final product, it will be necessary to test with a wider range of people to obtain really relevant and useful comments.

Ask the right questions: Each prototype should have some basic questions associated with it that you want answered. Before testing prototypes and recompiling feedback, you should be sure what exactly you are testing. For example, if you want to find out how easy a product is to use, you should organize the testing session to determine that characteristic and focus on discovering the positive and negative comments related to ease of use.

I like, I want and what if: This method invites the user to provide open comments with three types of statements:

- In the statements *I like...*, it is recommended to the user to transmit the aspects that he/she liked about the prototype. This provides them with positive feedback about your prototype.
- In the statements *I want...*, users are asked to share ideas on how the prototype can be changed or improved to address any concerns or problems.
- In the statements *what if...*, the user can express new suggestions that may not have a direct link to the prototype. This opens up possibilities for new ideas in future interactions.

Keep an open mind: Many times, testing sessions can reveal key points about previously unknown problems.

Remain neutral: We must be as objective as possible when presenting a prototype. Avoid highlighting the positive and negative aspects of the solution or trying to sell the idea. When people participating in the test express negative comments about the prototype, avoid defending it. Instead, do more research to find out what exactly is wrong. We must avoid being attached to our solution, and always be ready to change it, or even abandon it.

Adapt: When testing a prototype, we must adopt a flexible mind-set. For example, if we notice that certain parts of the prototype distract attention from its main functions, we should eliminate or change them to refocus attention on the key elements. In addition, if we see that the script planned for the test session does not work well, we should not hesitate to deviate from it and improvise to obtain the best possible feedback.

Comments box: A comment capture table (see Figure 7.2) is a structured way of organizing the comments that are compiled from the test sessions. A sheet of paper is divided into four quadrants:

1. If you label the upper left quadrant with the text "I like" or a "+" symbol: this is where the positive comments will be noted.

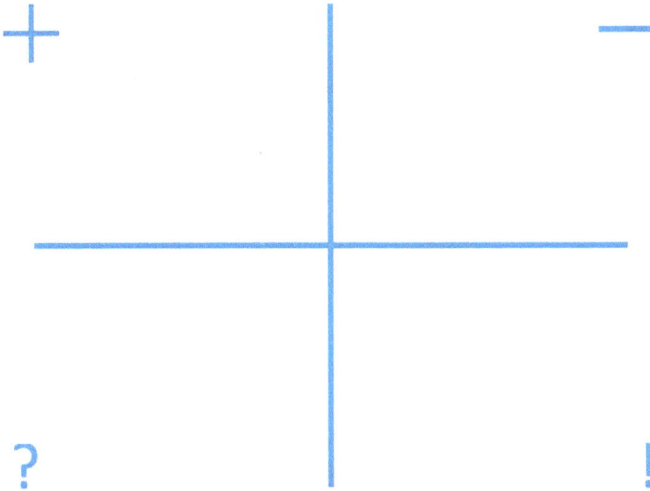

Figure 7.2   Comments box. The quadrants collect the positive comments (+), the criticisms (–), the users' questions (?) and the new ideas (!) generated during the prototype testing.

2.   The upper right quadrant is "Criticisms," where negative comments and criticisms about the prototype will be captured. We can also label it for example with a "–" symbol.
3.   In the lower left quadrant, there are "Questions." where you can write the questions asked by the people who participated in the test, as well as the new questions that arose in the test session. We can mark it with the sign "?"
4.   Finally, label the bottom right quadrant as "Ideas" or with the sign "!," and this is where you write down any new idea that provoked the test session.

Let the ideas flow: During the testing session, you should allow, and even encourage, the people participating in the testing session to come up with new ideas inspired by the prototype. For example, they can ask how the product or service could be improved. We can also reformulate some of the questions asked by the test subjects and be interested in their own opinion. For example, if someone asks about how to charge a new electronic device, we can turn the question around and be interested in what would be the best charging method for the new device.
Inspiring stories: Stories are powerful tools to inspire and think of solutions. In this process, one by one, all team members can share a couple of interesting and inspiring stories they observed when testing the prototype with users. When all participants are finished, they can examine the stories that were shared and look for common themes and possible ideas about the users to translate the inspiring stories into next steps for the project.

7.3   Useful tools

Next, we will describe some useful tools we can use in different steps of the DT methodology, including their descriptions, web pages or software to implement these solutions, and some interesting tips about how to use them.

7.3.1   Presentation techniques

Presentations are a fundamental tool within the DT methodology because they allow us to transmit our ideas in a structured and very visual way. They are a communication tool that adapts very well to many different contexts, for example, to show a prototype to a person or group of people, to communicate ideas, to give arguments for or against, as a support to give a speech, to distribute tasks among the members of a work team, to coordinate activities, etc. During a presentation, we transmit a message to our audience, and, in many cases, this message contains some persuasive element. It can be, for example, a talk about how well our team works, about what a candidate for a job position can offer to a potential employer, why my project should receive additional funding, or how ingenious our prototype is.

An effective presentation exploits (in a positive sense) the relationship between the person making the presentation and the audience. It takes into account the needs of the audience to capture their interest, facilitate their understanding, or inspire confidence. For our presentation to be more likely to succeed, good planning is essential. To begin with, we have to be very clear about what our objectives are, who our audience is, where we are going to make our presentation, and under what conditions we have to make it.

Some tools for making or sharing presentations:

- Prezi (http://prezi.com): It provides an online tool for making presentations in a very intuitive and visual way that requires little technical knowledge. It allows you to create presentations with animated transitions that unfold in a single space where the action takes place, moving from one element to another until a route is completed. A free version is available.
- Google Presentations (http://docs.google.com/presentation): Like the rest of Google's ecosystem of applications, this free application allows you to create and modify presentations, as well as collaborate in teams and teach them anywhere.
- Impress (http://openoffice.org, http://libreoffice.org): Part of OpenOffice and LibreOffice, perhaps the most popular open source office suite. It also supports PowerPoint file formats.
- PowerPoint (http://office.live.com): Probably the most well-known application for making presentations. Like the rest of the tools discussed here, the new versions also support online collaborative work and the distribution of presentations over the Internet. It also supports OpenOffice file formats.
- Keynote (http://www.apple.com/es/keynote/): Apple's presentation application for computers with MacOS operating systems, very popular among the users of these computers.

- Slideshare (http://slideshare.net): More than a tool to make presentations, it is a social network conceived as a platform to host presentations and share them in public or private.

In all cases, we can use these tools with our laptop or personal computer, and also with our mobile phone or tablet through the corresponding applications.

Objectives
Why are we making this presentation? We need to be very clear about what we want to achieve with the presentation, and what we want our audience to take away with them. What do we want our audience to assimilate with the presentation? What action do we want the audience to take after the presentation?

Once we are clear about this, we will be ready to make decisions about the design, style or tone of the presentation. For example, a presentation to sell our project may require a certain aggressiveness in the assessment of its benefits, and a presentation to apply for additional funding may require a more creative approach.

Audience
In general, audiences tend to be heterogeneous. Even if our presentation deals with a specific topic or a specific project, among our audience there will be people with different experiences, interests and levels of knowledge. To prepare a good presentation, we have to ask ourselves if our audience already has some knowledge or experience about what we are going to present.

Having a certain knowledge of our audience will also allow us to relate the content of our presentation to things they already understand or know, so that the speech will be more attractive and easier to assimilate. In addition, knowing how our audience breathes will give us clues about how easy or difficult it will be to convince them of our point of view.

It is not necessary to know each individual person in our audience, but it is essential to have general information about them to make sure that the material is appropriate. If we do not take into account the concerns and needs of our audience, it will be difficult for us to capture their interest or activate their imagination. For example, we should know our audience well enough to know if we should avoid or use technical terminology, if we should try to explain abstract concepts with practical examples, what effort we have to make to contextualize our ideas, etc.

Location
The scenario where we are going to make our presentation conditions to a great extent the way we relate to our audience. For example, a large theater can create a very formal atmosphere, while a small seminar can create a more informal atmosphere. Making a presentation from a raised platform can give an impression of distance from the audience, while proximity to our audience in a meeting room can encourage trust and participation. It is important to take into account if it is possible to modify the distribution of the furniture or the equipment of the place of the presentation, or even if we can opt for a place that adapts better to our needs.

Conditions

On some occasions, there are certain conditions that must be met in order to carry out our presentation. For example, we may be asked for a copy of the material to facilitate simultaneous translation or for later publication, so that we will have less room for improvisation. Other usual conditioning factors are the maximum time available, whether or not questions are allowed from the audience, if there is a maximum number of slides or a graphic style that we must follow.

Pecha-Kucha Presentations

A Pecha-Kucha presentation [58] is a type of presentation in which 20 slides are shown for 20 s each. Therefore, a presentation of this type always lasts 400 s, that is, 6 min and 40 s. With the Pecha–Kucha methodology, we achieve concise and agile presentations. In case our audience has to attend a session with several presentations (e.g., a proposal evaluation board, an award jury, potential clients of certain products or services, an audiovisual pitching, etc.), this methodology allows us to limit the duration of the sessions and provide all the presentations with homogeneous conditions.

Mastering the Pecha-Kucha methodology requires a certain amount of practice. Anyway, there are some ideas that we can take into account:

- Never forget that the conditions are strict, so organize well what you want to transmit taking them into account. For example, if the theme of your presentation can be organized in four ideas, dedicate to each five transparencies.
- Try that each transparency transmits a message in itself.
- Devise an outline for your presentation taking advantage of its format. For example, the outline could be a two-column, 10-row chart, where the first column is videos and the second column is text and/or images (see Figure 7.3).
- Try to be visual. Use images to increase the visual clarity of each transparency.
- Write a script for each presentation slide. Twenty seconds of narration should be between 30 and 60 words, depending on the speed of the speaker. Copy the script in the notes part of the presentation and/or in the outline.
- If you include text on the slides, this text should support and complement the narration, it should not compete with it for the audience's attention. The audience should listen, not read.
- Practice reading the script out loud until you are confident in your presentation. Do not forget that you have exactly 20 s for each transparency! Consider recording your presentation using the media provided by your preferred presentation application.

7.3.2 *Infographics*

The word infographics comes as a compound of the words "information" and "graphics." They represent a graphic and visual way to show information, data, or knowledge with the final objective that the information can be seen quickly and clearly. A typical

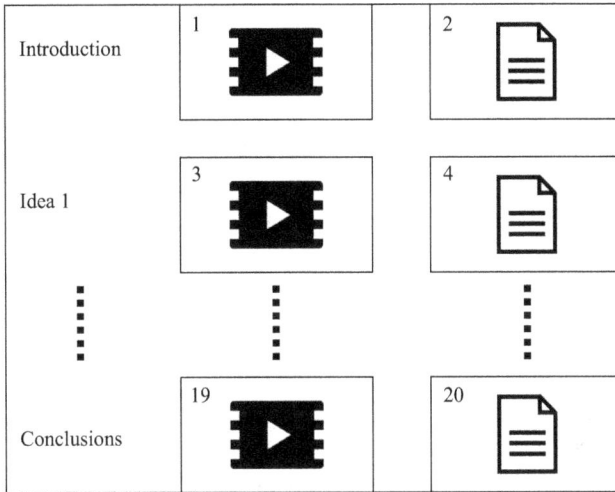

Figure 7.3   Example of Pecha-Kucha presentation

example of an infographic could be a metro subway map, where we can see graphically and quickly all the information related with the different lines, stops, connections, etc.

Infographics have existed for a long time, first known examples date from the seventeenth century. But are in recent years, with the extended use of the Internet, where they have become a mass communication method, tools like Adobe Flash in the early 2000s allowed us to create animations and present information as infographics. Nowadays we can create them using different tools that make use of HTML 5 and CSS3. Also infographics can be included in social media sites, like Twitter or Face-book, which has greatly increased their visibility around the world. But they are easily found in other types of media, like newspapers and television, to show the weather, maps with information, graphs with summaries of data, and a large number of other possibilities.

Parts

We can find three parts in any infographic*:

- The visual part relates with the colors and the graphics, and deals with the way to visually represent the data.
- The content part is usually based on statistics and facts that can be obtained from different data sources.
- The knowledge part is the insight contained into the data that the infographics are presenting.

*As seen in https://spyrestudios.com/the-anatomy-of-an-infographic-5-steps-to-create-a-powerful-visual/

Types

The more popular types of infographics are related to different ways to visually represent the data. Next we will briefly explain them:

- Time-series: is one of the most common forms of data visualization, where we usually present the time in the *x*-axis and the data to represent in the *y*-axis, and we can see the evolution of a set of values over time. Examples of time-series infographics are:
  - Index charts: mainly used when values are less important than relative changes, like with stock investors as they are less concerned with price and more concerned with the rate of growth.
  - Stacked graphs: represent area charts stacked on top of each other to represent aggregate patterns, allowing to see overall as well as individual patterns, but make them difficult to interpret trends.
  - Small multiples: an alternative to stacked ones, where instead of stacking the area chart, each series is individually shown, making each sector's overall trend easier to interpret.
  - Horizon graphs: they increase the data density of a time-series ones but preserving the resolution.
- Statistical distributions: are used to discover trends based on how numbers are distributed by means of calculations as frequency, mean, median, and outliers.
- Maps: are the natural way to represent data associated with its geographical location. In them, we can use different shapes, sizes, widths and colors to help encode information, or even create distortions in the shape of a region related with the data variable by redrawing them proportionally to the data.
- Hierarchies: many types of data can be organized following a hierarchy. For example, with node-link diagrams, following the structure of a tree, we can place the data into branches, allowing us the representation of multiple sub-sections.
- Networking: is used to graphically represent relationships, such as friendships. Examples of networking infographics are:
  - Force-directed layouts: where nodes are represented as repelling charged particles, with links that are used to join related nodes together.
  - Arc diagrams: are one-dimensional layouts of nodes linked with circular arcs those who are related.
  - Matrix views: have a node in each value in the matrix. By means of color, values associated with the links can be easily identified.

Tools

An infographic can be easily created using paper, ruler, pen, pencils and markers. But we have available different software or several online tools that can facilitate its creation. Here are some:

- venngage.com (https://venngage.com/blog/what-is-an-infographic/): this web page provides several examples of templates that can be used and even design tips that can help us create our own infographics.

- canva.com (https://www.canva.com/create/infographics/): this online tool allows us to select into a wide range of infographic templates, customizes it with images, illustrations and graphics, to finally saves it and shares it.
- infogram.com (https://infogram.com/): this webpage allows us to create engaging infographics and reports in minutes with the idea in mind of design with ease, enhanced with collaboration in real time, and with the final options to download, share and publish online.

7.3.3   Feedback capture grid

Already introduced as a tool on the section devoted to user feedback (cf. Section 7.2.3), the feedback capture grids are an amazing tool to record the information and feedback obtained in the testing sessions, allowing it to be presented in a structured way which will help us, as developers, understand what the users are trying to tell us. Also the information must be as detailed as possible. To do so we must encourage the users to explain all their information, feelings and every single detail they consider necessary about the prototypes.

Once the feedback capture grids are obtained, even if we have just one for all the users or one for each user, they will help us to find common ideas, demands or problems from the users, as well as to visualize their priorities. We can even cluster similar ideas or patterns inside them, giving us hints on how to solve the related problems.

How

Next we present a list of steps in order to use a feedback capture grid:

1. To start using one, we should divide a sheet of paper into four quadrants. Any surface to draw on will do, as a whiteboard, or an online document using one of the many feedback capturing grid templates that we can find online.
2. There is no general consensus of what to put in each quadrant, but one generically used could be:
   - Worked (or likes, +): on the top-left quadrant, we will write here the positive feedback, i.e. the things that have worked.
   - Change (or criticisms, –): on the top-right quadrant, we will capture here the things that must be changed, the negative feedback and the criticisms about the prototype.
   - Questions (or ?): on the bottom-left quadrant, we will write here to types of questions: (1) the ones that the users have asked and (2) the ones that the test session raised, the ones that we still have.
   - Ideas (or !): on the bottom-right quadrant, we will write here the ideas that the testing session with the users has sparked.
3. It is interesting to write in all the quadrants, at least a few notes. If you see that we are not obtaining enough feedback in one of the quadrants, it is recommended to guide the conversation with users to the corresponding topic of the quadrant.
4. Evaluate all the feedback capture grids obtained with your colleagues and discuss them in order to resume all the information and feedback received.

5. As we have emphasized during all this book, DT is not a linear path to find a solution to a problem. With all the information received in these previous steps of the feedback capture grid, you should go back to one of the previous DT phases in order to iterate to find a better solution that addresses the new information received during the testing phase.

7.3.4 Storytelling

Storytelling is yet another of the great tools that we can use in different phases of the DT process. In the testing phase, we can find use for it in two moments:

1. When introducing the users to the testing phase, we sometimes need to make them understand what they are testing, why they are testing it and where is the test taking place. All these can be done in a more appropriate way by using the storytelling tool, that just simply explaining the situation.
2. When we have done the interviews with the users and we need to share the data with the rest of the team not present in those interviews, an introduction using storytelling can facilitate the team to collaboratively understand the information and insights obtained in those interviews. In it, each team member present in the interviews shares their experience and observations in the interviews as a story, while the rest of the team take notes related to the stories.

Storytelling for the users testing

In the first case, we need to give information to the users that are going to test our products. We must make them feel informed, to know a little bit about the product, as well as comfortable, as if they were testing the prototypes in their location they are designed for.

One way to make the users feel involved in the testing is using a storyboard as visual illustration of the whole process for the users: a little bit of background, a short introduction to the prototype and a sense of information about the location where it is being tested. In this last case, if we have the chance to adapt our location to look like the real location where the product will be used, we should do it without thinking twice. Doing so, the users will feel more integrated in the whole testing process.

Then, continuing with the storytelling, the members of the group can use a little bit of role play, in order to act out the experience to the users that are going to be involved in the testing with the objective of giving them a better understanding of the prototype we are presenting. Apart from the role of the interviewer, we can think in a role of a presenter, which introduces the testing phase, a role of facilitator, which helps the users during the testing phase, and a role of observer, which watches the user experience and annotates feedback.

As the final phase of this storytelling, we will have the presentation of the proto-types, to which the users will arrive with a sense of immersion into the process that will improve and facilitate their experience. All this usually means for the team of developers to obtain more information about their prototypes in the interviews, which will lead to a better change to improve them.

Storytelling the interviews with the users
In the second case, storytelling works better just after the interviews with the users, when the memories of such activity are still fresh. This storytelling is going to be based on all the data collected during the interviews. The members of the team present in the interviews recounts each one of them, highlighting the observations and information that stood out. The rest of the team takes note of everything done. Once all interviews have been shared, the team gathers to share their notes and to find groups of them that have something in common. Once we have found those groups with things in common, it is necessary to identify a headline for each of them, as it should give us good insights of the information obtained in the interviews to the users.

Part II

Experiences of design thinking application in engineering

Chapter 8
Enhancing engineering by means of design thinking
Javier Yagüe[1], Manuel J. Fernández Iglesias[2] and Manuel Caeiro Rodríguez[2]

Design thinking (DT) is the central element of a long-term collaboration between University of Vigo and the Vigo Airport to introduce this methodology at their premises. DT is also the supporting methodology of a Master's level course on the societal implications of engineering taught at the University of Vigo. The course aims to analyze how engineering contributes to face the challenges encountered by modern societies and to provide the services they need. Enrolled students are expected to realize how the activity of an engineer is not an isolated endeavor with a limited impact, but it contributes to transform the world, both at a small and, eventually, at a large scale. Collaborating with the Vigo Airport provided an invaluable scenario, both to our students and airport staff, to address the problems and needs that airport users face, with the objective of providing the best possible airport experience.

8.1 Introduction

The role of engineering is to solve or mitigate problems, not to create them. Knowing how challenging situations were addressed and solved in the past can help to face problems in the future, which in turn leads to analyzing history oriented to future action, no to the contemplation of the past. Besides, engineering activities have direct influence on society, on how people live or how they relate to each other. In fact, most of the key societal changes in the last decades were based directly on contributions in the field of engineering. This requires that engineering students become aware of the ethical and societal implications of their work.

The University of Vigo established a long-term collaboration with the Aena-managed Vigo Airport in 2019. Aena, formerly Aena Aeropuertos, is a Spanish state-owned company managing general interest airports and heliports in Spain. Through its subsidiary company Aena Internacional, it also participates in the management

[1]Aena S.M.E., S.A., Public Business Entity in Charge of Airport Management Activity, Spain
[2]atlanTTic, Universidade de Vigo [GID DESIRE], Spain

of 15 additional airports worldwide. This collaboration enabled us to use the Vigo Airport as a controlled scenario to develop people-centered projects in the framework of this course. Also in 2019, Vigo Airport appointed the *Passenger Quality Committee* with the main objective of addressing the problems and needs of passengers who use the airport. The DT methodology offered the possibility for airport staff to connect directly with the passenger, to *wear their shoes* and overcome longstanding prejudices about passengers' needs and problems.

Although all airports appear to be the same in terms of their configuration (i.e., they all have an arrivals area, check-in area, departures area, boarding area, among other spaces), each airport has a different design and structure, and there are even differences in the services provided to passengers. The main reason for this is that the extension or the new construction of airport terminals is not carried out at the same time. This leads to each airport introducing a different design that in most cases is carried out by a different architecture studio. This implies that technologies, design guidelines and materials that are popular or fashionable at that time are used. As there is not common standard for airport design, this means that every airport has its own identity, even though passengers have the perception that all airports are basically the same.

However, in many cases, the companies and organizations that manage airports try to provide a standardized design and services, as many companies do in other sectors, by providing a common identity around the world. As a consequence, the sensation that a passenger perceives is that wherever they are in a location next to their home or thousands of kilometers away, they are aware that they are in a venue of a specific brand, thus creating a sense of common identity.

In the specific case of airport facilities, the question arises whether it would be a good idea for all airports to look the same in terms of design and services. Nevertheless, with the popularization of personalized services consequence of technological advances, a trend appeared towards the customization of the design and passenger services depending on the cultural environment or the type of traveler who frequently visits the airport. Besides, depending on their profile (e.g., leisure-biased or work-biased), passengers usually demand some specific services more than others.

The Master's programme in Telecommunication Engineering at the University of Vigo includes a course entitled *Telecommunications in the Information Society*. This course studies the societal implications of engineering activities and their evolution through history. For that, students solve problems or situations identified by them after their interaction with real people in controlled scenarios. Thus, they should apply their social abilities, soft competences and engineering skills. By applying the DT methodology, challenges would be identified within some field of study or environment, and all the information available related to that challenge would be gathered. Students, working in groups, would propose imaginative solutions and select one of them to construct a prototype to be tested with real people. The solution achieved will have to consider not only technical questions but also legal, environmental, social and those related with sustainability.

Our collaboration with the Vigo Airport provided an invaluable setting for bringing the activities in this course to life, that is, for our students to propose and develop people-centered projects addressing the problems and needs from airport passengers, with the main objective of providing the best experience possible. At the same time, the application of the DT methodology in airports is perceived as necessary and essential by Aena, due to the fact that Spanish airports are firmly committed to improving the passenger's perceived quality, a perception that a priori seems simple but, being subjective, depends on the type of airport user, as well as the cultural environment in which it is located, its weather characteristics or its rural or urban nature, among many other characteristics.

The next sections discuss the activities developed in the framework of the agreement between the University of Vigo and Aena to use of Vigo Airport as an experimentation space for the students enrolled in course *Telecommunications in the Information Society*. After analyzing the introduction of DT at the Vigo Airport, the course aims and the suitability of DT as the methodology of choice are introduced in Section 8.3. Then, we showcase a selection of student projects carried out in the framework of our collaboration with Aena. The assessment instruments utilized both within the course and at the airport are also briefly described. Finally, we offer some conclusions and identify some lessons learnt.

8.2 The introduction of design thinking at Vigo airport

In April 2019, the Vigo Airport's passenger quality committee was created with the motto *If you were a passenger…What would you improve in our airport?* This committee was made up of workers and staff from different departments at the airport. Eventually, a multidisciplinary group of 14 people was appointed with different skills and perspectives about the airport. None of the committee members knew anything about the DT methodology, so the first challenge was to introduce this methodology as an inspiring source of innovation. Besides, DT was used as a reference to establishing the committee's dynamics, objectives and commitments, with the support of the lecturing staff of *Telecommunications in the Information Society*.

Eventually, the commitments below were made:

- To be curious and observant. Any piece of information provided by a passenger is important and may provide a source for inspiration.
- To be empathetic, both with passengers and fellow workers and with their circumstances. Being able to put ourselves in the other's shoes.
- To question the *status quo* and avoid prejudices or assumptions.
- To be optimistic and positive, losing the fear of making mistakes, and seeing mistakes as opportunities.

It was the first time that most of the quality committee members were part of a disrupting project like this. The commitments made were inspired in by the DT methodology, and more specifically by the attitude of Design Thinkers when addressing a project's initial empathy phase. Indeed, DT would serve as a guide for any activity carried out by the committee from its inception to its final implementation and testing.

8.2.1   *Methodological approach*

The first committee meeting was devoted to lay the methodological foundations to address the challenge of improving airport services inspired by actual passenger needs. In the case of the empathy phase, several instruments and tools were analyzed to assess their suitability for empathizing with airport users. Eventually, it was decided to use the tools below:

- Empathetic interview. Carrying out face-to-face interviews with the passenger in different areas of the airport.
- Shadowing and observation of passengers within their context. The main processes in an airport were identified for monitoring: check-in, departure area and boarding, arrivals, baggage claim and transportation from and to the airport.
- Research on how other airports are addressing this challenge. Benchmarking of those initiatives from passengers' perceptions.
- Online research about the interactions in social media and other sites of Vigo Airport users: opinions, ratings, discussions, etc.

These tools would also be used by the student groups addressing challenges at the Vigo Airport.

Committee members were distributed into two groups. The first group would be dedicated to conducting interviews and shadowing. The other would be in charge of benchmarking tasks and online research to collect information and opinions from Vigo Airport users. As a preliminary approach, three basic questions were posed to gain initial insight about passengers' perceptions:

- What do you like the most about our airport?
- What do you like the least about our airport?
- What is missing in this airport?

The intention was to discard any existing preconception and gather fresh information about passengers' opinions, perceptions and needs. The simplicity of the questions promoted honest, sometimes candid, answers. Answers to more specific questions at this stage could be conditioned by existing prejudices or assumptions, both from committee members and respondents.

The team carrying out these interviews in the field was given some basic but important advice, to avoid as much as possible the influence of the prejudices that staff accumulates over years of experience. Besides, due to a sociocultural issue, the character of the Galician passenger when answering questions is hermetic, ambiguous and sometimes imprecise. This requires an added effort on the part of the interviewer. The interview guidelines below were provided both to quality committee members and student groups:

- Work in pairs and carry out shadowing and interview at the same time. Take notes.
- Ask the reason for the answers provided, do not presuppose. Ask and delve more into the type of answer given by the passenger, with the idea of obtaining as many information as possible, as detailed as possible.

- Pay attention to non-verbal language and give the passenger enough time to express themselves confidently.

On the other hand, and introspection exercise was proposed for all committee members. They would ask themselves about how they can make passengers at Vigo Airport feel more comfortable, safe, reduce that fear, stress, apathy, restlessness, uncertainty, when taking a flight.

8.2.2   Discovering the actual needs of airport users

Vigo airport can be considered a medium-sized regional airport with its own type of passenger profile, namely a business passenger due to being located in an industrial area with strong international connections. However, in recent years the vacation passenger profile increased considerably, reaching 50% of all airport users. The Vigo area and the Rias Baixas region also have their own identity. Both airport's and geographical identities would be included among the innovation ingredients. By addressing passengers' needs and desires, they would remember our airport as a singular entity, being aware of their visit to Vigo and the Rias Baixas and recalling it in the future as a pleasant experience, even an unforgettable one.

The information capture process took 2 months. On June 2019, a second meeting was organized to put together all the information gathered. The group in charge of shadowing and interviews summarized their findings in relation to the three questions posed. When passengers were asked about what they like the most about Vigo Airport, they stressed the comfort and size of its only terminal, which was perceived as accessible requiring short walking distances to access any service offered; the efficiency of airport processes, requiring short waiting times; the kindness of the staff; the proximity to the city center and the general comfort and quietness of the airport.

When questioned about what they liked least, passengers pointed out the few international connections; the quality and scarcity of signage; the high prices of some services such as parking or catering; the need for more charging stations for mobile devices and the limited schedules and destinations of local public transport. Something that many passengers missed, which on the other hand would not be difficult to address, is a means to check the time anywhere around the terminal.

Regarding missing services, apart from the ones mentioned above, passengers pointed out the lack of business or VIP lounges; signage and support to visually impaired users; playgrounds for children; additional shops and catering services and a clearly identified passenger information point.

The group researching passengers' perceptions on the web detected that the most common complaint was related to the inconveniences of passport control and security filters. In some cases, security measures being less strict in other airports (e.g., not having to show some small electronic appliances packed in hand luggage) generated negative comments and ratings.

Other negative comments were a consequence of one of the weaknesses identified during the interviews, namely the poor signage. Some passengers complained about the lack of diaper changing stations in restrooms for people with disabilities. This is not the case, as there are two changing areas for people with disabilities, one in the boarding area and another one in the public concourse. However, these changers are not signaled as such.

Some comments confirmed other weaknesses. For example, the lack of some services available at larger airports, poor communications by public transport or the lack of a VIP lounge contributed to a substandard passenger experience in some cases. Two recurring complains are the small number of destinations and the extremely high prices of catering facilities in relation to perceived quality.

On the positive side, most passengers are quite satisfied with the comfort, the lack of crowds, the tranquility and the ease of access from the city center by public transport. Finally, in latter reviews and comments passengers highlighted the airport's network connectivity and the introduction of charging stations for small electronic equipment.

Information was also compiled about what passengers valued most in major international airports around the world. In general, results were in line with Vigo Airport passengers' opinions and perceptions: the cleanliness of restrooms and common areas; a varied gastronomic and commercial offer at a reasonable price; the availability of comfortable areas to rest or work while waiting for a flight; the availability of playgrounds for kids and leisure areas for adults, or the airport having good signage and being easy to navigate.

All information gathering activities above demonstrate that services and facilities beyond those related to airport operations are most relevant for a positive passengers' experience. In fact, these elements contribute both to users' satisfaction and to enhance an airport's identity. For example, some airports integrate airport services with original elements to create unique airport experiences, such as a local brewery, a seasonal market, native art exhibitions, and gardens with local vegetation or animals.

The insight gained as a result of the information gathering activities above inspired some project ideas intended to provide a solution to key passenger needs. These ideas were discussed with airport users by means of additional interviews. Passengers would be introduced to the new projects, questioned about what they think about them and given the opportunity to contribute their own ideas, improvements, and new proposals and suggestions.

- What do you think of the idea of enabling a multicultural space with exhibitions, music, dance in the Terminal?
- What do you think of the idea of placing a piano in the terminal so that random people could share their playing abilities, as well as the idea of organizing musical events?
- What do you think of the idea of setting up green areas with trees, plants and seats around them? For example, putting olive trees in the terminal or some other type of local tree or plant that could grow indoors.
- What do you think of the idea of putting up a video wall, a giant screen in the public arrivals area? The idea is to show cultural events in the area, as well as information on airport destinations and tourist information on the Rãas Baixas.

All the proposals above correspond to projects beyond standard airport services that were intended to contribute to enhance the user experience.

8.2.3   Time for action: shaping the traveler-centered Vigo Airport

The results obtained by the quality committee both motivated and justified the implementation of several intervention projects in the airport facilities along the next years. The next paragraphs discuss the most relevant.

As pointed out above in the previous section, many passengers demanded an easy way to check the time. The group eventually decided to integrate a clock at the bottom of the flight information screens scattered all around the airport. Thus, passengers have available in a comfortable way and on the same screen, all relevant information for traveling, such as flight schedules for departures and arrivals together with the current date and time. The cost of this measure was very limited and passenger feedback was very positive.

The group also addressed the deployment of a second airport store at the boarding area, at that time with no commercial offer whatsoever. The new store would be located at the busiest boarding area. This new commercial space would offer passengers gastronomic products (preserves, chocolates, local confectionery), Galician handicrafts, premium wines and spirits from Galicia and Spain, educational toys, jewelry and souvenirs. The new store also offers tax-free products to passengers from countries outside the European Union.

The silent airport concept was also implemented taking advantage of the existing passengers' perception about the airport's quietness. Boarding calls are no longer announced throughout the public address system and the rest of announcements were limited to those considered essential. This reduced environmental noise and effectively made the airport quieter.

In January 2020, the new airport VIP lounge was inaugurated (cf. Figure 8.1). This lounge was given its own personality and renamed Sala Illas Cies. Illas Cies (Cies Islands) is a major international touristic destination in the Vigo area, ranking among the top 100 sustainable destinations in the world.

Figure 8.1   New VIP lounge at the Vigo Airport

The airport's security checkpoint was also refurbished in 2020. Not only were the facility renovated, but it was also made more spacious and comfortable for the passenger, enabling an exclusive access area for families with children and for persons with reduced mobility.

An Aena customer service and information office was also opened in the main concourse. Besides airport staff at office operating hours, a free video call service system was also deployed. With this, passengers may contact and be assisted by Aena staff at any time. This service point is adapted for people with reduced mobility.

In June 2021, an integral refurbishment of the restrooms was carried out. All-automatic toilets were installed to ensure they remain clean between service times. Signage was also updated and improved with colorful and illuminated pictograms. The updated restrooms' signage is part of a more ambitious project addressing the improvement of signage across the Vigo Airport.

Finally, a working and a resting space were deployed at the boarding area. These spaces were custom designed for the Vigo Airport. The working area was equipped with 10 workstations, 2 of them adapted for people with reduced mobility, while the resting space was equipped with 10 comfortable seats and 230 V and USB fast charging outlets. These spaces are located in a bright space, close to the most used boarding gates and very close to the restaurant area, with a flight information point right opposite. Space was reserved for live plants and other ornaments, such as an olive tree, the symbol of the city of Vigo. This intervention will be done in the framework of a comprehensive decoration project for the terminal building, with native trees and plants. This project was inspired by the Changi airport in Singapore, as well as by the feedback obtained from the passenger interviews.

8.3   A DT-enhanced course on engineering and society

Telecommunications in the Information Society seeks to motivate students in the application of the technical concepts of engineering to solve problems and offers services in the society in which they live. Students are intended to become aware that engineering as a discipline does not have an isolated technological impact, but it contributes to transform the world, both on a small and a large scale. This insight leads to two fundamental consequences:

1. Society, the people around us, have problems that can be solved by engineering professionals. The aim of engineering is to solve or mitigate societal problems, not to create them. Knowing how other challenges were addressed in the past can help to face future problems, which in turn leads engineering professionals to study history from the perspective of future action, not as a means of contemplating the past.
2. Engineering activities have a direct influence on society itself, on how people live and how individuals relate to each other. In fact, most of the dramatic societal changes in recent decades were directly driven by contributions from engineering activity (e.g., transport technologies, biotechnologies, radio

communications, television, mobile telephony, the Internet, etc.). This influence must be accompanied by a commitment to ethical responsibility.

Students have to learn how to apply technical knowledge considering the social and ethical implications linked to the application of that technical knowledge. They have to understand the complexity and the consequences of making technical decisions based on incomplete or limited information about their social implications. In practice, this means to perform technical activities considering the safety of people and goods, to develop the ability to understand the ethical responsibilities of their work and to understand and observe the applicable laws, regulations and standards.

Besides, engineering professionals have to develop skills to integrate information & communication technologies and systems in broader and multidisciplinary contexts, such as bioengineering, health and medicine, nanotechnology, economics, environmental sciences, agriculture, and transport.

Finally, students have to become aware of the need for lifelong training and continuous quality improvement, showing a flexible, open and ethical attitude toward diverse opinions or situations, in particular, in terms of non-discrimination by sex, race or religion, respect for fundamental rights, accessibility, sustainability, etc.

To sum up, after completing the course, students will have a solid knowledge about what being an engineering professional means and what it represents, together with full awareness of the social, ethical and environmental responsibilities of engineering as a discipline. They will also be enriched from the contact with other disciplines in which technology is integrated to drive societal change.

8.3.1   Course organization

Having in mind the aims above, the course is organized in three blocks:

Engineering & Society Seminar: This seminar is devoted to illustrate the professional activity of telecommunication engineers and its ethical implications through invited talks from active professionals, some of them graduates from our school. Students are invited to interact with speakers. The seminar also introduces DT as a methodology that encourages future engineers to observe society and try to find solutions to challenges that directly and personally affect specific individuals or groups of people. Airport staff is invited to participate in this seminar in the framework of our long-term collaboration with Aena.

History of Telecommunication Engineering: This is a lecture series devoted to the historical evolution of the telecommunication engineering profession in the fields of line communications, radio communications, technical consulting and major services such as television, the Internet and mobile telephony.

A Multidisciplinary Society: Students work in groups of 4–6 people to address a challenge using DT. Proposed challenges are based on societal problems or situations not strictly related to telecommunications engineering, so that students understand their role in a multidisciplinary endeavor and how they can contribute by applying their engineering skills and abilities. The Vigo Airport serves as a real-world scenario to address some of these challenges.

As pointed out above, project in the Multidisciplinary Society block is addressed by means of a project-based learning methodology, and more specifically DT. Student teams address the resolution of a specific case or project, from its definition according to the early DT stages to the final presentation of the outcomes in a testing session with the participation of lecturers, invited professionals and the rest of the class. Students have to document their activities in writing and develop a presentation for the testing session that may include any of the approaches introduced for the DT prototyping stage.

Student teams will be configured according to the outcomes of a simple personality test to be administered at the first lab session. The objective is to distribute students into heterogeneous groups, mimicking the way project teams are configured in a real-world labor setting.

Course projects are interrelated and are framed within a specific environment or application field (e.g., a commercial center, the airport, the COVID-19 pandemic, etc.). However, students are not assigned a completely defined project, but they have to identify situations or challenges in the defined environment, not necessarily related to telecommunication engineering, that need to be addressed and solved. For that, they have to apply the DT methodology, and more specifically what they learned about the empathy and definition phases. When working towards a solution, they will have to address not only technical issues but also legal, environmental, social and sustainability-related issues should be considered. Students will eventually come up with imaginative solutions and will try to develop a proposal that reasonably addresses the problem identified, although it may not yet be implementable given current technological development. The final aim of this experience is not about manufacturing or programming a market-ready solution, but to provide a socially acceptable proposal that addresses both the technical and non-technical aspects enumerated above.

Teamwork begins by identifying and locating all the relevant information. From this information, they will try to identify the people involved and will try to empathize with them, to state the problem they feel and not the one students think that they identified from the outside. Once the problem to be addressed is explicitly stated (i.e., defined according to DT standards), groups will devise technological or procedural solutions to that problem. They will have to find the technical and scientific information required to produce a prototype, report their findings and the overall process, and develop a presentation to be used together with the prototype in the final testing session. Students are proposed to document their activities by means of an online service such as a forum or a wiki.

The interaction with lecturers will take place in five 1-h sessions distributed along the academic semester and through an online forum. As project milestones, student teams will submit the identified point of view (PoV) at the definition phase by the date of the second session, and three tentative ideas to solve the problem by the date of the third session.

8.4   Student projects at the airport

Along the next paragraphs we discuss some of the projects carried out in the framework of our collaboration with Aena. The DT methodology would allow students to abstract themselves from existing prejudices and start from scratch, to understand what is truly relevant for the passenger who frequents Vigo Airport.

As a common starting point for all airport projects developed in the context of this course, students would put themselves in the passenger's place to identify and understand their needs, adding value in an authentic and innovative way. With this, it would be possible to gain distance from passengers' stereotypes and delve into what the passenger really cares about, their needs and desires, to eventually design a rewarding experience for all passengers when visiting the airport premises.

Innovation plays an important role in this process, since technology is the basic tool utilized to improve airport processes. Technology, together with innovation, would allow us to gradually reduce and eventually eliminate those negative prejudices that passengers have about airports.

8.4.1   Enhanced airport trolley

Moving around a large airport terminal can be expensive or overwhelming for people who does not travel regularly or people with special needs, such as people with disabilities or seniors. Besides, health-related restrictions consequences of the COVID-19 pandemic are especially relevant in all situations that involve physical presence. Access restrictions were broadly enforced according to the size and capacity of physical spaces open to the general public, such as airport facilities. In this context, this project seeks to help an individual who is not used to traveling by plane to move around in a comfortable and safe way, paying special attention to the specific needs of senior travelers.

In this case, design thinkers paid several visits to the local airport to interact with travelers of all ages and airport personnel. In the case of travelers, they focused on non-regulars, that is, people traveling for the first time or persons traveling only sporadically. In the case of airport personnel, our students tried to find out what were, according to their experience, the more demanding situations encountered by this kind of airport users. In both cases, the tool used was the empathetic interview. Travelers were addressed while waiting at the airport. Delayed flights happened to be a great opportunity to experiment this tool, as the persons to be interviewed had a plenty of time available and do not have the pressure to complete other activities previously planned. Interviews with airport personnel were scheduled in advance according to their availability.

The crafted PoV that summarized the empathy findings and completed the definition phase was as follows:

> Maria, a senior traveler visiting her grandchildren abroad, needs to have clear
> references and feel familiar with airport facilities, because otherwise the

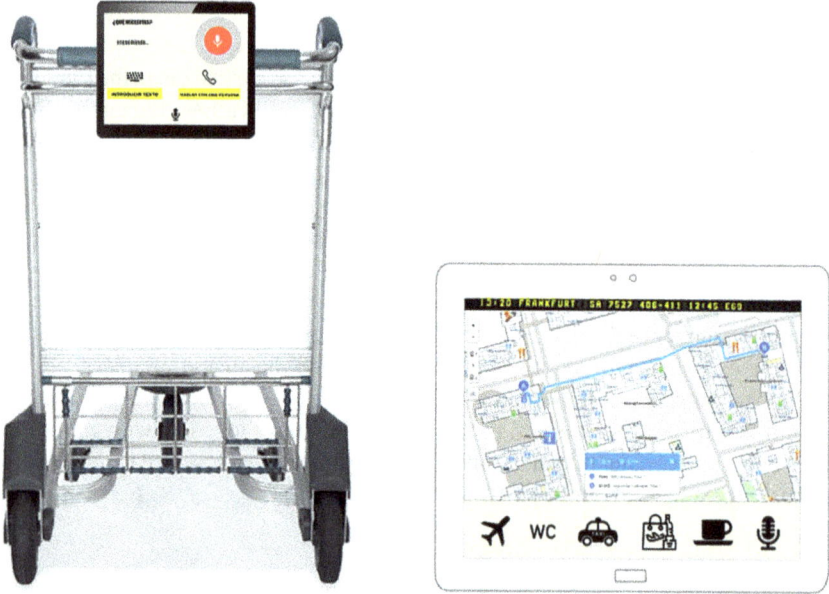

Figure 8.2   Mock-up of the trolleys' application. Basically, airport trolleys are equipped with a tablet computer hosting the airport app.

airport becomes a hostile environment that makes her not feel like traveling again in the future.

After a brainstorming process and the classification or ideas by means of a *Now, How? Wow!* diagram, the design team came up with a technical solution consisting of an integrated application on a tablet, which is attached to regular airport trolleys (cf. Figure 8.2). The application scans the passenger's ticket and will immediately compute the most comfortable route to reach the departure terminal and gate. Machine learning techniques will be introduced to optimize route calculation taking into account passengers' feedback and the outcomes of previous routes proposed.

The tablet attached to the trolley will display a map of the terminal clearly identifying the passenger's location and their destination, together with relevant airport services such (e.g., restrooms, information points, cafeterias, restaurants, etc.). The application will support user interaction by means of voice or text commands for passengers to select a new destination within the terminal. Destination can be selected from a structured drop-down menu, by a name or a brand or by expressing a need for some type of service or assistance. For example, the following interactions below will eventually provide a route to the next vending machine selling bottled water:

1. Selecting *Services>Catering>Soft drinks & water* from the drop-down menu.
2. Responding with *Water vending machine* to the prompt *What do you need?*
3. Responding with *a glass of water* or *water*.

A mock-up prototype was developed to test the solution. The prototype consisted of several sheets of paper representing common passenger interactions attached to a tablet-sized board, which in turn was attached to a trolley. Actual airport passengers were invited to try the prototype to obtain feedback about its features. The solution was also introduced to airport staff.

Passengers' perceptions about the enhanced trolley were very positive. The most valued feature was the possibility of knowing anytime and anywhere around the airport their exact location and distance with respect to the boarding point. More than half of the passengers trying the prototype indicated that they would use such an app if it were available for regular smartphones. However, they also valued in general this service being provided by the airport.

On the other side, airport staff pointed out that the logistic requirements of deploying such a service were difficult to overcome, no matter that they also valued the solution's concept. Most relevant concerns were related to tablets' maintenance (e.g., charging stations and charging cycles), the security risks of managing a large network of wireless devices in airport premises and passengers' data protection issues.

8.4.2   Overcoming corona fear

The COVID-19 pandemic hit our lives and radically changed them. Practically all places where human interaction occurs had to adapt to guarantee that they were perceived as safe places. To do this, measures were implemented such as prescribing a safety distance among people, the introduction of face masks in certain areas, enforcing hand hygiene with hydro-alcoholic gels, etc.

One of the social activities most affected by this situation was air travel. At the beginning of the pandemics, air travel was drastically limited. Even today, travel restrictions are in place and many people still perceive airports as public infrastructures where chances of getting infected increase, no matter this perception of augmented risk is not supported by evidence in most cases. To avoid this situation, airports should prove to their users that their facilities are safe places to enjoy, even in case of long stopovers.

Thus, design thinkers were proposed the challenge of designing and providing airports with a solution with which they can ensure that travelers who have to make a long stopover feel safe while waiting for their next flight.

The empathy phase took place at the local airport premises. As in the previous case, airport users were interviewed taking advantage of the opportunity provided by idle times consequence of flight delays or early arrivals to the airport. Passengers and other airport users (e.g., people picking up or dropping off passengers, service personnel at airport facilities, family members) were questioned about their perceptions of the safety of airport facilities in the COVID-19 context. The most common concern was related to the cleanliness of facilities, especially those used intensively by many people such as restrooms. They also would greatly appreciate if the airport provided up-to-date information that would allow them to avoid crowded areas as much as possible.

All findings were crafted in the PoV below:

> Juan and Leonor, a couple traveling with their kids, needs to feel safe while in the airport, because any incident involving them or their family may compromise their return trip home, with the corresponding expenses and other tribulations.

Eventually, as the most promising outcome from a structured brainstorming process, designers proposed as their solution a mobile app to be used inside airport facilities with the functionalities below:

• An up-to-date map of the airport divided into zones, which can be selected and enlarged to visualize the number of people who visited that zone in given time range. When the number of people exceeds a configurable threshold, the area will be marked in red. If the number of people is close to the security threshold, the area will be marked orange. Finally, if the number of people does not pose a perceived danger to the traveler, the area will be marked in green.
• It will allow the traveler to know the current occupancy level in each area, so that they can decide if they want to visit that area or not.
• The user will be able to know when service areas were cleaned up, either in person by using the different NFC tags available in those areas, or by selecting in the app the control points that appear in each area. The app will provide the same information as scanning the NFC tag.
• Peak occupation times will be predicted to help the user decide when to attend any of the service areas. For this, flight schedules will be used and a machine learning model will be trained from previously recorded data on the number of people in different time ranges.
• Restrooms will be equipped with capacity counters including a dual sensor, one outside and another inside each location, to verify whether a person actually enters, approach the door but do not enter, or leave. A notification will be sent to the cleaning service when the actual occupancy exceeds a threshold value. In addition, a display at the entrance will indicate the number of people currently inside and the number of people that have visited the premises since the last cleaning session.
• It will allow passengers to book slots at common service areas that may have occupancy restrictions at certain times. For example, airport users would be able to plan when to go to a restaurant, clothing store, etc. knowing that the maximum regulated capacity will not be exceeded.

A wireframe mock-up (cf. Figure 8.3) was developed to present the solution concept and its basic features to airport passengers. The mock-up prototype included all relevant layouts and functions for discussion and feedback. Besides, a short commercial was developed to promote the new application by means of an animated slide presentation. The prototype served to confirm that the features included in the

Figure 8.3 Mock-up of CovidAir application. Each zone map provides visual information about occupancy and cleaning status. Blue squares represent NFC tags signaling cleaning status.

application were positively valued and to introduce some layout modifications to enhance users' experience.

A feasibility study was also carried out to confirm that the proposal was technically feasible, since it involved known technologies with which the design team already worked thanks to other courses in their previous bachelor's degree in telecommunication engineering and present in master's.

With this project, a promising solution was proposed to tackle a challenging air travel-related problem, such as user safety in airport premises, which requires offering services to passengers so they are perceived as safe. A solution for common service areas was designed, where all airport users typically converge, so that they can know in advance which areas are safer than others, being able to choose which one to use. They can also find out which areas were just cleaned or even reserve a space for personal use in public capacity-restricted service areas, such as restaurants or airport stores.

8.4.3 Secure luggage claim

The airplane became over the years a means of transport for the masses. More and more passengers decide to fly to reach their destination and, consequently, more and more luggage pieces are at airports. This increased people and baggage traffic eventually results in a less satisfying user experience. The waits, the crowds and the concern that someone may misuse or take your belongings generate insecurity and mistrust among travelers.

No matter the security enhancements that major airports around the world implemented along the last decades; in many cases, baggage collection is still a process largely based on trusting other fellow passengers. In many situations, especially on domestic flights, passengers just collect the piece or pieces of luggage identified as their own, and proceed to the airport exit. No matter technical advances, baggage claim is still perceived as a slow, obscure and distressing procedure that contributes with additional discomfort to an already distressing experience such as air travel.

Design thinkers were challenged to enhance the process of collecting luggage upon arrival to increase its security and to reduce the possibility of another passenger taking a piece of luggage that does not belong to them. The main problems that travelers face upon arrival at a new airport are associated to baggage claim. In this process, there are frequent delays and the consequent crowds of people waiting for their luggage to be deposited in the arrival carousels. Likewise, as pointed out above, and as a consequence of the enormous relevance of the human factor in this procedure, it is probable and highly undesirable that a user's luggage is mistakenly picked up by another traveler.

As their first task, the design team analyzed the existing airport baggage handling system at the local airport. They spent time observing travelers collecting their luggage and interview luggage handling staff. They discovered that airports today already have relatively complex baggage handling mechanisms. These systems are made up of an interconnected network of conveyor belts and destination-coded vehicles (DCV) that link all the airport gates and connect them to each other. The objectives of these systems are as follows:

1. Move luggage from check-in to the boarding gate.
2. Move luggage between two different boarding gates (on scales).
3. Move luggage from the arrival gate to the baggage claim area.

Although our area of interest is focused on the third objective, it is interesting to review how objectives 1 and 2 are addressed. When a passenger delivers their luggage at check-in, a label with a barcode is attached upon entering the baggage handling system. Then, a 360-degree barcode reader scans this code. Once scanned, baggage pieces are located at all times and a series of sorting machines direct them to the departure gate. At the gate, operators load the luggage in a cart or in containers to be loaded in the plane. This last step is still fully manual in most cases.

Upon arrival at the destination airport, the process is usually simpler since all the bags on the same flight are already classified and clustered together. They are unloaded from the plane by baggage operators and transported to the designated baggage belt. From there, a series of sorting machines direct them to the designated baggage claim carousel.

The PoV that would drive the ideation phase was as follows:

Drew, a frequent flyer saleswoman, needs expedited and undisturbed access to her luggage upon arrival to her destination, as it includes her working gear and the items needed to make her feel comfortable while doing her job abroad.

With the information gathered at the airport, three different approaches to address this challenge were analyzed: an on-demand luggage collection system, an evolution of existing luggage carousels and a shift-based system. On-demand luggage collection, as opposed to traditional luggage collection from conveyor belts, is based on travelers explicitly claiming their luggage. A passenger, after disembarking from the aircraft, goes to the luggage claiming area and uses their baggage tag to collect a QR code from a teller machine, either printed on a piece of paper or electronically sent to their smartphone. With that code, the traveler is responsible for claiming their luggage at their best convenience by scanning. Upon scanning the QR code, the passenger be assigned a free baggage collection slot or cubicle from a pool of collection slots. In parallel, baggage pieces are scanned and sorted to be delivered to the destination slot when the QR code is scanned. Delivery is supported by an automated conveyor belt-based classification system conceptually similar to parcel classification systems at postal facilities.

The second approach is based on the modification of existing airport conveyor belts. A series of closed boxes would be installed between them and the passengers where luggage pieces will be deposited. In addition, the belt should be closed to prevent anyone from directly accessing luggage. Mechanical arms would be installed along the belt for pushing the luggage pieces to the deposit through a padded slide, avoiding possible damage to the luggage. Upon scanning the baggage tag at the baggage room, the passenger is provided with a box identification for the box where their luggage will be delivered. When the system detects the arrival of a user, their luggage, which is rotating on the belt, is identified and put in the designated box by a robotic arm. Then, the passenger can open the box using one of their baggage tags.

Finally, the third approach is inspired by the systems utilized in some restaurants to avoid waiting lines, where patrons waiting for a table are notified by means of a wireless device or directly to their mobile phones. For example, a mobile application could be used to notify passengers when their luggage is deposited on the conveyor belt at the luggage claim area. The system may also utilize the passenger's phone number provided upon reservation or check-in to notify them by means of a short text message.

The three proposals were discussed with frequent air travelers at the university, researchers and lecturers that regularly attend international conferences or research facilities abroad. The most popular approach was the on-demand luggage collection system introduced above. Thus, this option was selected for further investigation and ultimately for prototype development.

This new baggage claim system adapts the existing conveyor belts by installing lockers around them, to which luggage pieces would be deposited. Each locker would remain continuously closed, except for the moment in which the traveler who owns the luggage pieces inside unlocks it. Barcodes in baggage tags and 360° scanners would be used to assign baggage pieces to a specific locker.

The system orchestrates baggage collection according to passengers' demands. Once at the destination airport, travelers have two options to claim their luggage: through a mobile application or through a physical teller machine. Using a mobile application, a traveler would be able to scan or import their luggage tag. The baggage claim request will be associated with the generation of a code, immediately available to the user. When the corresponding luggage pieces are deposited in one of the lockers on the conveyor belt, a notification will be sent to the passenger's personal device, so that they can go to their locker and retrieve their belongings safely. Likewise, arrival notices will be announced through visual panels throughout the airport.

Users who do not have the mobile application would proceed to a teller machine in the baggage claim area to scan their boarding pass or baggage tag. Then, a paper slip with a QR-code and a text tag will be printed to be picked up by the traveler. When their luggage is deposited in a locker, a notification with the tag and the final pick-up locker will be announced through the airport panels, at which time the passenger will be able to retrieve their baggage from the locker indicated.

The system may also offer different qualities of service. For example, the free basic option may assign lockers in a first come first served basis, while a premium option would assign higher priority and to access a free locker. Besides, some lockers and teller machines may be allocated to the premium service only to further expedite luggage claim.

Note that this solution does not require any structural modification related to baggage check-in.

8.5 Assessment

Students' assessment aims to find out whether students are able to integrate knowledge from different sources and take evidence-based decisions, even when information available is incomplete or limited. Besides, they should demonstrate that they understand the ethical responsibilities of the engineering procession, and show an attitude towards the sustainability, accessibility and quality of engineering outcomes. Finally, they should acquire abilities to apply telecommunication and information technologies and solutions in multidisciplinary contexts, such as health sciences, social sciences, and arts and humanities. Different assessment methodologies are utilized according to each course block.

Engineering & Society Seminar: Students have to complete a questionnaire for each talk with questions about the impact that the activity discussed may have on their future careers, on their perception of the engineering profession and on society in general. Students' interaction with the speaker is also considered for grading. This block contributes 20% to the total course grade.

History of Telecommunication Engineering: Students have to write an original essay about one of the technologies introduced along the course. The essay must contribute a critical view about the reasons and motivations why the selected technology was introduced, its societal impact and its role in the evolution of telecommunication technologies and in technological progress in general. This essay contributes 15% to the total course grade.

A Multidisciplinary Society: Assessment is organized in two phases. First, student groups have to write a memo describing the project's journey and how the design thinking methodology was applied along the different project phases, which contributes 25% to the total course grade. Finally, student groups have to make a public presentation of their projects and their outcomes. The presentation will be evaluated by both students and faculty using evaluation rubrics similar to those introduced in Chapter 11. Individual students' and lecturers' scores are averaged within each group, and both groups contribute equally to the final grade, namely 20% each.

On the other side, the instrument utilized by airport staff to check whether DT produced the expected results is the monthly surveys in the Airport Council International's (ACI) Airport Service Quality (ASQ) program carried out at the airport. These surveys provide a subjective assessment from passengers about many different aspects at the airport. An analysis of the 3-year period 2019–2021 was carried out, being 2019 the year in which the Quality Committee initiated its activities. Year 2019 was compared with 2020, the year in which some of the ideas began to materialize as a result of the DT initiative. Year 2020 was compared with 2021, the year in which most DT projects initiated by the Quality Committee discussed above were executed. In 2022, a large part of the ideas will be developed and implemented, always driven by the insights captured from actual passengers. It can be observed an increase in the general satisfaction of the airport passenger in 2020 compared to 2019, which is even greater in 2021 compared to 2020.

8.6   Conclusion & lessons learnt

From the lecturers' PoV, we believe that with the introduction of the DT methodology the learning objectives and outcomes of the course are fully met. The main aim of *Telecommunications in Information Society* is to make students aware of the social and ethical implications of the engineering profession. In this context, the introduction of a methodology that puts the actual people benefiting from the engineering work at the center of the process has a dramatic impact on the students' perception about the impact of their decisions. With DT, they discover that technical decisions not only contribute to improve people's lives through innovation, but also influences people's perceptions about technological advances and their role in shaping society.

This course also promotes collaborative problem solving. By means of course projects, students train relevant soft skills, such as critical thinking, listening to others, asking better questions, generating ideas, creating better stories or inspiring and sustaining collective action. Besides, from a more technical perspective, DT fits perfectly with state-of-the-art agile software development methodologies. In these, projects are carried out by means of short cycles of functionality development where the presence of end users is strongly encouraged.

The introduction of the DT methodology in this course brought a significant improvement over more traditional project management methodologies introduced in similar courses. This improvement is evident both in the development and achievement of course objectives and in the students' degree of involvement. Annual academic satisfaction surveys consistently show a very positive attitude of the students toward the course. Their subjective perception regarding the learning outcomes and the quality of the work carried out is also very positive. Furthermore, the solutions and prototypes crafted by project groups effectively respond to the real needs of the target users, which contributes to increases their satisfaction with the work carried out. Students realize the key role of the empathy phase, of interacting with real people, to banish preconceived ideas that they may acquire from working exclusively with the information available about the problem at hand. They also value the path through the definition and ideation phases to correctly frame and interpret a problem as a means to eventually identify the most appropriate solution. These phases are perceived as very visual and dynamic, very far from the educational methodologies common to most of their previous courses. From the lecturers' perspective, it is important in these phases to support different visions and ideas from students, so that they realize that solutions are not unique and that there are different ways of approaching a real problem. Finally, they value very positively the possibility of constructing a prototype in a very short time, as prototypes provide a sense of usefulness to the work that students are doing.

The collaboration with Aena facilitated the introduction of DT at the Vigo Airport, which in turn motivated airport staff to connect more with the passenger. In turn, passengers found a way to transmit to airport staff their perceptions about the airport. This methodology allowed both staff and students to approach, from a different perspective, the problems that actual passengers have, overcoming prejudices and interpretations coming from passengers' opinions received through official suggestions and complaints which, in most cases, do not meet the most appropriate conditions for the passenger to calmly expose a problematic situation.

The airport will continue to develop and materialize the ideas generated into actual projects driven by the DT methodology. In short, this project can be considered a most enriching initiative, which will allow all stakeholders to keep contributing with innovative ideas to adapt and improve a key transport infrastructure in the Rias Baixas region, so that visitors would ultimately live a great experience that motivates them to come back in the future.

Chapter 9

New automotive engineering proposals based on design thinking

Katarzyna Znajdek[1] and Anna Laska-Leśniewicz[1]

This chapter presents the results of the experiment that was conducted in order to face one of the challenges in modern automotive engineering – autonomically driving cars. Design thinking (DT) methodology was used to check the needs of the users, dispel their doubts and find the solution to make them comfortable and alleviate their fears about this new and controversial feature in automotive industry. The project was undertaken with the support of the companies, simultaneously by several various groups.

Automotive branch is a demanding area for designers and engineers. DT can effectively support improvements and solutions applied in future cars. An interesting result from the experiment was that in spite of the fact that each team was working in a very similar environment, including the same geographical conditions, they all presented different solutions considering the varied needs of the end users. For the purpose of the process analysis in this chapter, we will present the path undertaken by one of the groups and the final results proposed by the team within the project. The team consisted of three international students who were working on the topic of "autonomously driving vehicles" at Lodz University of Technology for one semester.

9.1 Introduction

The transportation sector is undergoing several profound changes as there is rapidly advancing technology. Demand and users' expectations are adjusting to the developing world and available possibilities. According to the report prepared by a World Bank Group team, five major innovations may be distinguished [60]:

1. Oncoming sharing-economy and sharing platforms in transport and mobility.
2. The rapid improvement of electric batteries and the development of the other alternative fuels for motor vehicles.
3. The development in machine-learning techniques associated with big data that enable real-time information processing.

---

[1]Łódź University of Technology, Poland

4. The growth of eCommerce and express door-to-door delivery of goods and services.
5. The improvements in autonomous vehicles (AVs).

Autonomous driving technology has been developed intensively in the last few years. An autonomous vehicle is defined by National Highway Traffic Safety Administration (NHTSA) as [61,62]: "A vehicle whose operation takes place without direct intervention by the driver to control steering, acceleration, and braking and which is designed in such a way that does not expect to constantly check the road, when the automatic mode is running."

Fully-autonomous cars will shake current transportation systems and revolutionize society [63]. There is a wide range of outcomes currently being forecasted and they can be varied from place to place [64]. In [65], authors present the possibility to do other tasks than driving, which can appeal mainly to drivers who will "gain" additional time. Moreover, many researchers indicate that shared self-driving vehicles will be embraced to a large extent and public transportation will diametrically change itself. Users will prefer to use shared AVs offered by ride-source companies similar to Uber rather than possess their own cars [66]. Even if many various scenarios are debated, the majority of experts agree on the fact that understanding public attitude towards technology and innovation is an important starting point when designing and engineering future mobility and autonomous vehicles. Public acceptance must be obtained and some social benefits are highly welcome [67,68].

The demand for mobility of people and goods in urban areas has significantly grown and is estimated to increase further [69]. Increasing the infrastructure will not satisfy this tremendous growth, as according to [70,71] in 2050, two-thirds of the world population will live in urban areas and the total amount of distances travelled within urban zones will triple compared to the nowadays situation. Technology-based innovations that can answer to quick and radical changes in the world seem to be necessary and very beneficial. Therefore, autonomous vehicles are one of the interests for mobility, similar to artificial intelligence, electric mobility, big data and data analytics, Internet of Things (IoT) and Internet of Everything (IoE), 5G and connected vehicles (V2X), and technologies for Blockchain transactions [69]. Autonomous vehicles seem to have many advantages, among them, the most important are the following ones:

- significant reduction in the number of traffic accidents – and increase in road safety;
- optimization of traffic flows and better environmental impact;
- reduction of traffic congestion;
- mobility available to the entire population regardless of age and place of living (complete territorial accessibility);
- reduction in resource consumption and environmental pollution;
- reduction of parking areas;
- transformation of time spent driving from unproductive to productive;
- decrease in "driver costs" as shared mobility will be available and cheaper than a currently used human-driving model.

LEVEL 0	LEVEL 1	LEVEL 2	LEVEL 3	LEVEL 4	LEVEL 5
*No automation (manual control)	*Driver assistance	*Partial automation (vehicle can accelerate, brake, control direction)	*Conditional automation	*High automation	*Full automation

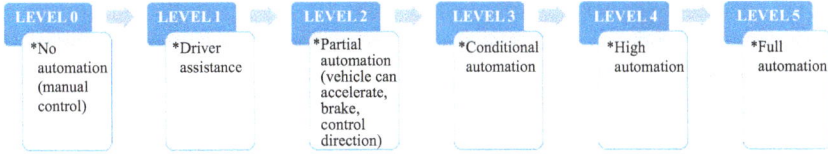

Figure 9.1   SAE levels of driving automation (graphic based on SAE International's standards for driving automation levels)

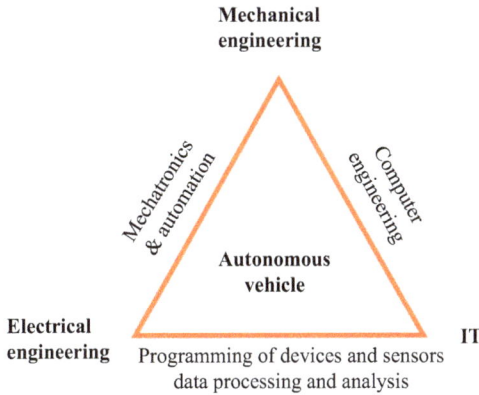

Mechanical engineering

Mechatronics & automation

Computer engineering

Autonomous vehicle

Electrical engineering   IT

Programming of devices and sensors data processing and analysis

Figure 9.2   Main disciplines for future mobility – autonomous vehicles

The degree of autonomy of the car can be verified using different classifications and standards coexisting with each other. The most adopted and followed by the scientific literature are the standard published by the NHTSA and the standard published by the SAE (a standardization body in the field of the automotive industry) [72]. The SAE J3016 standard is often treated as the reference standard. It has established six levels of autonomous driving that are based on the greater or lesser degree of automation of the vehicle, with the relative level of human participation in driving the car. The SAE levels of driving automation are shown in Figure 9.1. Current commercial autonomous vehicles (model Tesla Model-s/X, Volvo XC90, Audi A8) are classified at L2 and L3 levels.

In order to achieve a certain level of autonomy, the car exploits the ability to detect the surrounding environment through techniques such as radar, LIDAR, GPS, and sensors. Therefore, the interaction between these components and the advanced control systems on board the car allows the latter to make decisions about the paths to follow and any obstacles and signals to monitor. From the engineering point of view (PoV), innovative solutions in future mobility and autonomous vehicles require the collaboration between at least three disciplines – electrical engineering, mechanical engineering, and IT (cf. Figure 9.2).

DT methodology has been applied in many fields, especially in business and IT for over two decades. Automotive industry looks for innovations as well. New solutions should combine three main areas – feasibility (technology), desirability (human-oriented design), and viability (business aspects). DT allows designers to work and create in interdisciplinary teams, including real users and experts at every single design step, and underline the importance of iteration to achieve great success. For autonomous vehicles, DT may help to get balance between rapidly developing technology and real human needs. Even if automotive engineering is full of novelty, the DT approach changes the perspective of designers and gets them closer to people that have also some limitations and psychological barriers in contact with technological transformation.

9.2   Empathic research, users and defined needs

The first task accomplished by the students, in order to approach the issue of autonomous vehicles, was an inside-group analysis of the possible advantages and disadvantages concerning undertaken topic. Such analysis was considered beneficial for framing the topic and understanding how individuals could interface with each of considered problems.

The first discussed positive aspect of autonomically driving cars was the issue of safety. Studies show that death by road accident is the main cause of death among young people in Europe and is mostly linked to mistakes, distractions or reckless actions, rather than to real unexpected events that are difficult to predict. Therefore, a machine designed to follow the rules is expected to be able to overcome this problem.

Moreover, autonomous cars could represent an important change especially in lives of people with disabilities or those who have mobility difficulties, since they would be able to regain a good level of autonomy. Students recognized the fact that this technology is not only aimed at those who does not want to drive but also to those who unintentionally cannot be drivers.

Furthermore, in students' estimation, autonomically driving cars, which represent the market of new technologies, have some major benefits for the industry and society, such as progress in human achievements, scientific development, establishment of new companies and new job positions. Economic compensations can be located in the reduction of fuel waste while driving manually.

Final advantage defined by the group can be, in their opinion, the engine of this change. This is the peoples' time and stress optimization, specifically, the possibility of doing other things while traveling and avoiding situations often considered stressful, such as driving in traffic or looking for a parking place.

Accompanying this series of advantages, students also defined significant disadvantages, which are related more to the difficulties that must be faced to achieve this change, rather than to the real negative aspects, at least for the most part. First issue is the fact that there are places in the world that are not particularly suitable for the circulation of cars, and also for this reason it could happen that car journeys not only between countries but also between cities are more difficult due to different regulations between the place of departure and arrival.

Additionally, it appears evident that a period of coexistence between autonomous and manual driving cars is inevitable. This can lead to difficulties in the safety aspect of autonomous cars, since the human variable is not be removed from the equation, even when the first fully autonomous cars begin to be present on the road. In that case, there is also a question of whether it is actually worth it. The technological, bureaucratic and economic challenge is evident, which arises the concerns of directing the efforts elsewhere.

Students conclude with more subjective aspects, however, equally important. In fact, there is a large number of people who like to drive, it is something they do with pleasure and they may not be so willing to give it up easily, or they prefer to be responsible for their lives having full control over the wheel.

In the end phase of this internal discussion, a series of more delicate doubts and ethical questions appeared. First, there is the question of the job issue. It is considered to be true that this technology would open new doors for new jobs, nonetheless it can seriously reduce other professions, such as drivers, mechanics, and school drivers instructor. The second question considers peoples' privacy. In autonomous driving system, the entire route and the current position of the car are indicated and processed by the software. Finally, there are some moral question to be addressed. The responsibility in the event of an unexpected, emergency situation, where a sudden maneuver is to be undertaken or taking risks endangering the life of passengers.

9.2.1   Research on automotive market

Automotive market can be recognized as the most innovative one. Research and development are the areas in which a vast amount of money is invested in that industry [73]. Customers seem to be at the center of the improvements in delivered products. At least, products (i.e. cars) must be attractive from the customer's PoV, the industry wants to create the need of following new automotive trends, which leads to a high sale rate. Automotive industry has changed over the last 3 years. The main reason is a pandemic of COVID-19 around the world.

Until early 2020, automotive industry showed a constant increase in terms of sales. This tendency was observed for both, fuel vehicles and electric ones. While the market share of the EV was constantly growing, the interest in cars with diesel engines was dropping significantly. This was due to the bad reputation of diesel engines after the diesel gate scandal, as well as legal regulations introduced in many European cities aimed at limiting the zone available for internal combustion vehicles (especially diesel vehicles).

The beginning of 2020 and pandemic disrupted the automotive market like many other areas of the economy. Sudden lockdowns and production stoppages led to the problem with the availability of the vehicles. Global logistic chains were broken due to the pandemic, which resulted in the shortage of components for the new vehicles. The most significant example is semiconductors that were highlighted as one of the most crucial components that were missing. After the first lockdowns during the year 2020 and 2021, the situation was slowly coming back to normal. However, the demand was still higher than production capacity. It was the time when lots of small manufacturers (especially from the far East) introduced their concept of a small electric car that can be affordable (sale price lower than USD 20k). Big manufacturing concerns were focused to stabilize the production and introduce to the market new models (mostly electric but also conventional ones). The beginning of 2022 and the crisis related to the war in Ukraine creates a new wave of disruptions. Suddenly it turned out that almost all car manufacturers either produce their vehicles in Ukraine and/or Russia, or at least some components are produced there. The results of such a situation especially after the pandemic discrepancies have led to the sudden stop in the production and have created even higher gap between the market demand and the production capacity. Causing an increase in the waiting time for a new vehicle even up to a year! Some manufacturers adjusted the list of equipment to eliminate some components that were produced in the impacted areas. Some of them decided to focus only on a few models and stop the production of others. In both cases, the result has a significant impact on the customer and their buying decision. In most the cases, a future owner does not decide what car to buy taking into consideration his/her preferences or price, but the decision can be made based on the availability of the vehicle.

Another aspect is related to the energy crisis caused by the war and the fact that most European countries depend on Russian oil and gas. As a result, prices of energy suddenly have increased and the distribution has become more a political than economical issue. It drives the increase of the prices on the gas stations and strengthens, even more, the tendency of looking for the EV vehicles or at least hybrid ones, which might be a noticeable saving for the user.

The current situation is very dynamic and it is hard to predict how the future will look like, especially with the latest announcement from the EU to ban the sale of new petrol vehicles since 2035. Based on the available data, global electric car sales are growing and this trend will expand. In 2021, the total market share was equal to 6.0%. Detailed data published by Automotive Research Company – JATO is presented in Figure 9.3 [74]. In August 2021, following models entered the ranging of TOP10 of EV vehicles sold on the market: Volkswagen ID.3, Tesla Model 3, Volkswagen ID.4, Renault Zoe, Ford Mustang Mach-e, Kia Niro, Skoda Enyaq, Tesla Model Y, Volkswagen Up, and Fiat 500 [75]. The market is still developing.

9.2.2   Interactions with potential users

After recognizing the automotive industry market, students defined several groups of stakeholders, in order to address the issues connected with the topic of autonomous vehicles in everyday life.

The architecture of the considered groups, including their placement on the graph representing relationship of the decision power and the level of the relevant interest in change, is shown in Figure 9.4. More detailed explanation of the stakeholders and their roles in this analysis is given subsequently.

In Figure 9.4, the team selected, grouped and described several of the above-mentioned groups of interest.

- **Potential customers**, including:
  – people who are sensitive to environmental protection,
  – those who care about savings,

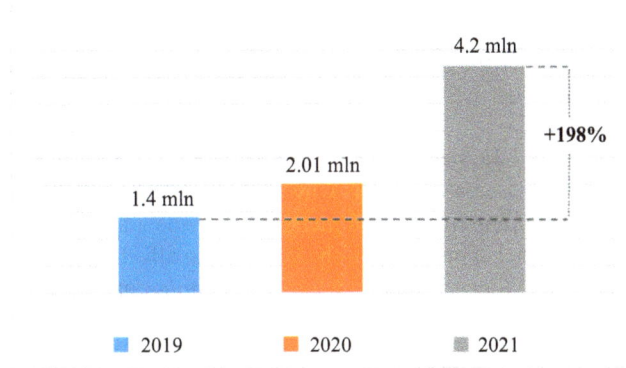

Figure 9.3   Global electric car sales in 2019, 2020 and 2021

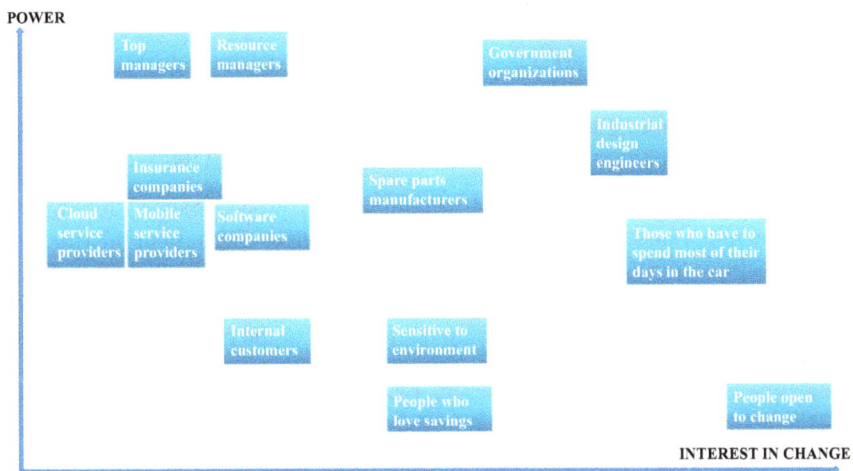

Figure 9.4   Stakeholders map (based on students' design)

– people open to change,
– people who wants innovation every day,
– those who have to spend most of their days in the car,
– people who like to be intertwined with technology.
• **Spare parts manufacturers**: This, according to students, is one of the strongest and most important stakeholder group. Spare parts production is essentially important for the maintenance of cars and other electro-mechanical vehicles.
• **Government organizations**: It is an inevitable end that the government of each country has the knowledge of every innovation. For this reason, the needs of the country and the requirements for such new solutions must be defined and regulated. This group is also responsible for ethical issues and law regulations.

- **Software companies**: The biggest envisaged problem with the software of the vehicles is the reliability issue. The license and reliability of each mobile application purchased for the software used in the vehicles should be an important and considered issue, which makes the software companies another important stakeholder group.
- **Industrial design engineers**: Vehicle designs can vary according to the needs and preferences of a person. That is why a stakeholder group which aims at the aesthetic understanding of the customers will provide a significant value in the process. The external appearance of the vehicle can affect the financial success.
- **Cloud service providers**: Data protection is very important. In order not to store all the data in the memory of the vehicle, which can be very dangerous, all user data can be collected in a cloud system. In order to provide security and assurance of data privacy, students selected cloud service providers as another important stakeholder group.
- **Insurance companies**: Most of the parts or vehicles in automotive industry have either a guarantee or insurance. Insurance companies are an important player in automatization process at this market.
- **Mobile service providers**: One of the indispensable issues of autonomous vehicles is constant mobile service. Customers can for example expect the car to drive from a parking lot to the given location using mobile phone, computer, or smartwatch. Mobile service providers are essential to fulfill this request.

Stakeholder groups selection and definition in DT process must be followed by conducting interviews with several of their representatives. Because of the final task to find a valid solution for certain needs, students needed to understand these people thoughts, problems and desires, as potential users of this future technology. This task was very difficult and time consuming, as some of the groups listed above are extremely hard to make contact.

Initially, the team members interviewed a group of their friends, who were mostly university colleagues and former schoolmates, who share similar interest in science and technology. As a result of this part, students received a vision which was generally congruous and consistent with their internal considerations. However, according to the interviewees of this group, they noted minor importance of any privacy risks and low focus on the issue of the necessary coexistence period of autonomous and manual cars.

In further steps of the investigation, the project team contacted with some experts, with whom the interview took more technical turn and was less tied to personal opinions. In particular, students for example obtained the confirmation that most road accidents are caused by human errors. One of the experts claimed that studies have yet to be carried out to establish with certainty the best level of autonomy of the cars. Additional suggestion was that there must not necessarily be that the fully automatic solution is the best.

9.2.3   Recognized problems, needs and challenges to be met

It is significant that the designing team deeply understands the problems, need and challenges faced by the user of the final outcome. DT is a human-centered process,

Empathy Map

SEE
✓ When I can use the vehicle in my environment, I can get excited and stressed
✓ After using the product, I can feel very comfortable using it, also I think I can be happy
✓ Invasive risks to our privacy already today

THINK & FEEL
✓ Worried
✓ Unsure
✓ Hopeful for advantages
✓ No more traffic monsters
✓ Thinks fights in traffic will decrease
✓ Thinks that time spent in traffic can be used better
✓ Thinks people with disabilities can drive and socialize
✓ Thinks people's lives can be saved faster thanks to autonomous vehicles- automated cars are safer
✓ Coexistence is necessary

SAY & DO
✓ I wouldn't want to have a car that I can't drive
✓ A self-learning tool can destroy my control - I do not want this
✓ I want to try the product, but I'm afraid of the cost
✓ I believe in future technologies
✓ I want to leave a better world to my daughter

HEAR
✓ People will probably be surprised
✓ Many people would not accept sudden changes
✓ I don't understand how this works
✓ Cars used to be better

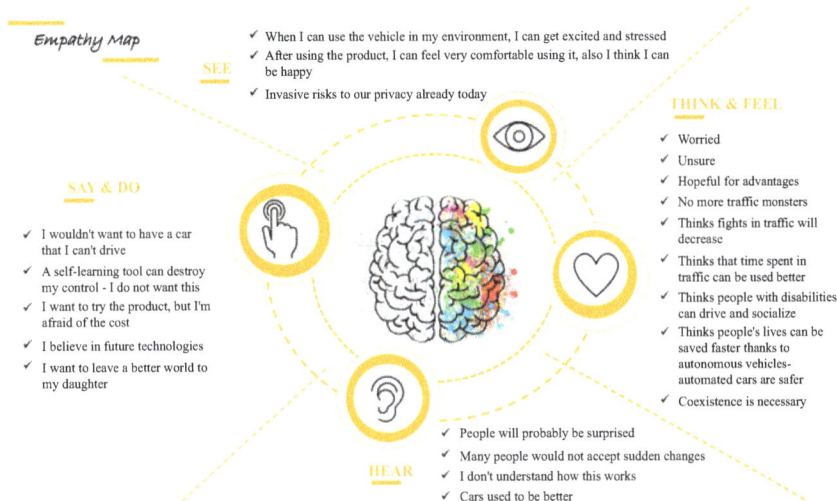

Figure 9.5   Empathy map (based on work designed by student Fatma Vural)

thus the individual perspective is crucial for a successful effect of the project. At this point, students made an effort to condense the information from the previous steps and then grouped into various fields on the empathy map, which is given in Figure 9.5.

Summing up the problematic and challenging issues, according to the interviewed potential users of autonomous vehicles, the group of students focused on what they found the most common in most of the answers.

Autonomous cars generally make people feel safer. Safety has always been one of the first things people mentioned when questioned about this topic. In addition to this, students were told about other, less important but still present advantages, such as the possibility of avoiding stressful situations of driving in traffic or finding parking place. Despite that, however, most people believe that a period of coexistence between a system composed of both autonomous and manual cars is necessary because of the people's habits. They find it difficult to imagine that we will suddenly switch from one system to another, even because there are some critical aspects to consider.

Some people have also expressed doubts about the possible situation of driving a car without knowing how to drive it and most of them would still want to have their driving license, not giving a total faith towards this technology. Furthermore, there are remaining doubts concerning the costs of this technology and its implementation on the market, including not only cars' adjustments but also roads' infrastructure standardization.

Most interviewees, therefore, claimed that taking all the aspects into consideration, they look enthusiastically forward this new technology. Students' team declared that there was the majority of the positives expressions, even despite the awareness of efforts that would be required to achieve that result. Consequently, it is also possible to consider this as a collective effort to make the world a better place for future generations.

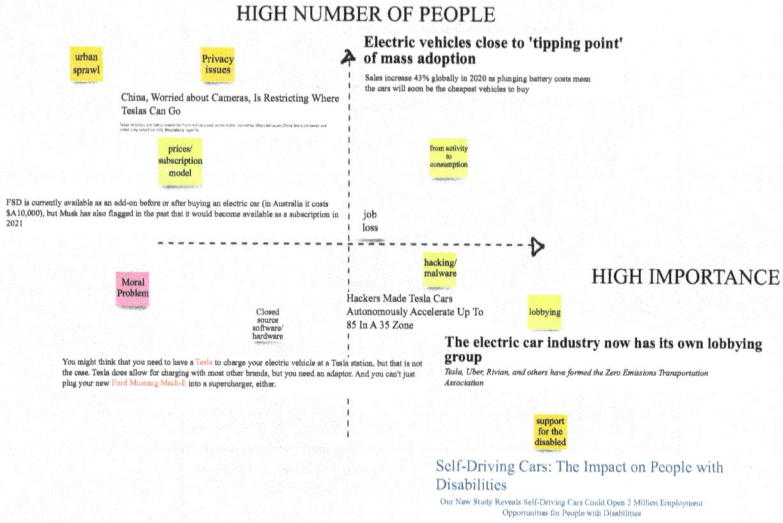

Figure 9.6   Problem mapping graph (created by students Lorenzo Bruno, Ababacar Diouf and Fatma Vural)

Real problems are extremely complex entities, as humans are infinitely faceted and cannot really be described in simple terms. The main approach that opens the way to a new "rational" understanding of reality is the invention of Cartesian coordinates, giving order by affirming that something is more important than something else. Thus, applying this technique to the given challenge, the project team built a number of maps, enabling them to frame the problem in analytical terms. One of them is shown in Figure 9.6. It is based on two parameters, the importance of the problem and the number of people involved. The map was created in order to find the hot spot for the investigated idea.

9.3   Target group of automatic cars users and their PoV

In order to fully define the target group of end users, students needed to find common issues in personalities, desires, problems and interests of the people that they selected to design for. For this purpose, they described and named the representative, combining all the above-mentioned characteristics, in one picture of the persona.

The activity of creating persona consists of the conception of fictional, but, at the same time, realistic profile of the defined target customer, in order to truly address his/her needs. Persona diagramming is a deeper way to understand end users of the designing product, compared to traditional research and demographic methods. It allows to create more personalized strategies, therefore it positively impacts the customer experience. The group applied this knowledge to project needs mixing multiple inputs. The persona of the project is presented in Figure 9.7.

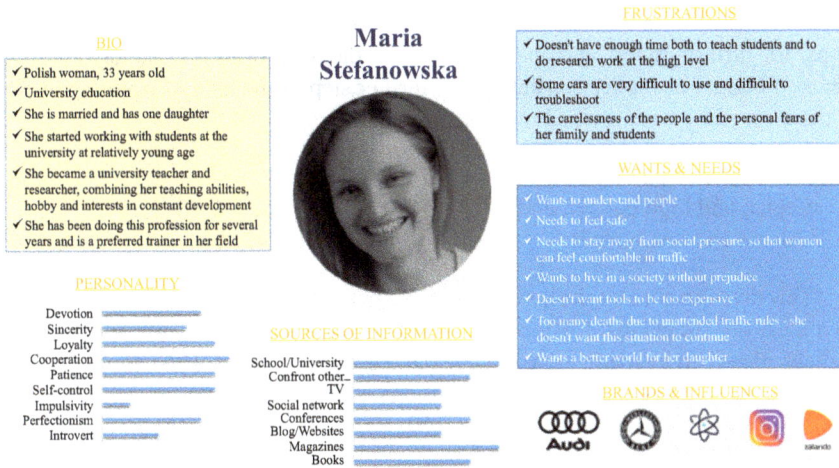

BIO

Maria Stefanowska

FRUSTRATIONS

✓ Polish woman, 33 years old
✓ University education
✓ She is married and has one daughter
✓ She started working with students at the university at relatively young age
✓ She became a university teacher and researcher, combining her teaching abilities, hobby and interests in constant development
✓ She has been doing this profession for several years and is a preferred trainer in her field

✓ Doesn't have enough time both to teach students and to do research work at the high level
✓ Some cars are very difficult to use and difficult to troubleshoot
✓ The carelessness of the people and the personal fears of her family and students

WANTS & NEEDS

✓ Wants to understand people
✓ Needs to feel safe
✓ Needs to stay away from social pressure, so that women can feel comfortable in traffic
✓ Wants to live in a society without prejudice
✓ Doesn't want tools to be too expensive
✓ Too many deaths due to inattended traffic rules - she doesn't want this situation to continue
✓ Wants a better world for her daughter

PERSONALITY

Devotion
Sincerity
Loyalty
Cooperation
Patience
Self-control
Impulsivity
Perfectionism
Introvert

SOURCES OF INFORMATION

School/University
Confront other
TV
Social network
Conferences
Blog/Websites
Magazines
Books

BRANDS & INFLUENCES

Figure 9.7   Persona of the project (based on the data prepared by students)

Created persona is a middle-class, educated and financially independent woman who has a family consisting of a husband and one kid. She is influenced by technology, news in science and social media. Her main focus is on safety, stability and self-development. She wants to be up to date with innovations and technical gadgets; however, she is not willing to spend too much money on the newest tools. She is aware of the changes in the world and wants to be a good example for her daughter and her students.

An integral part of the DT process is the definition of a meaningful and actionable problem statement, on solving which the designers want to focus. This PoV statement is essential for deep understanding of specific users, their needs, and the most essential insights about them. Defining PoV is defining a real problem to be solved. This definition is not obvious and it needs mindfulness taking in consideration previous findings. However, this process enables designers to find surprising insights about their users, in order to apply them while proposing the solution. In this case, the PoV statement is as follows: an adult independent woman, who is a full time teaching and research university employee, needs to be relaxed to recover from a busy workload, because relaxation increases the speed of finding solutions to problems and improves general focus of everyday activities. From this point, the group of students started practicing creative ideation focused strictly on meeting this particular need of the defined persona user.

9.4   Creative ideation on future innovations in automotive

The process of creative ideation is one of the students' favorite steps in DT methodology. The brainstorming process produces a high number of ideas, therefore they are needed to be arranged in a meaningful way. This was accomplished by the young designers by using affinity mapping, as a technique allowing a large number of ideas to be proceeded from brainstorming to sorted groups, based on their natural relationships, for further review and analysis. At first, the ideation was conducted with no limits of budget or reality constraints. In further step, to be able of better analyzing of the ideas, they were divided into three groups:

- controversial and original,
- utopian and high budget, and
- realistic and low budget.

This division enables designers to better focus strengths and weaknesses of their own ideas as, for example, Utopian ideas may be adapted to real-life situations, as realistic ideas can be made more complex and groundbreaking. The list of generated solutions together with their short descriptions is presented below.

Controversial ideas

- System of interconnected automated cars. This is a very futuristic project, that might happen when the period of coexistence between automatic and manual cars is over and all cars are autonomous. It requires the removal of all manual driving cars and vehicles.
- Reserved, specific areas of the city in which there is a concentration on constructive efforts for the disabled people. This solution can consume a lot of state funds, and, moreover, people may not be in favor of moving away.
- Providing automated car services for the disabled. This addresses the issue that automated cars are not only useful for people that do not want to drive a car but also for people who cannot do it. The autonomous cars can represent a very important change in their lives.
- Workweek reduction. This is an idea that has been repeatedly considered and it is also already implemented at some circumstances. The advantages from a social PoV are evident, with more time available to spend with loved ones; however, according to some studies, it can also be advantageous from a working PoV by improving the quality of the work done by people who are more rested and motivated.

Original ideas

- Headphones for dogs guides. Headphones which are connected to a GPS signal, for example from the phone, send simple voice commands to the dog, such as "right," "left," so that the dog can lead the owner to the place where the car was parked. It obviously requires city construction to take this idea into consideration.
- Intelligent system of automatic telephone calls. Autonomous car capable of understanding, through an internal database of movies and other information, the type of emergency that is occurring, and consequently alert the competent authorities.
- Sounds capturing system. Noise can be a warning of imminent danger and, in this way, such possibility is also given to the people with poor hearing. Sounds are transmitted into the car with a sophisticated 360-degree audio system, so that even the direction of danger is distinguishable.

- Smart glasses. Smart glasses that are able to show the way to the chosen destination, all equipped with a microphone for voice commands. Such glasses must have a minimal design, for reasons of comfort.
- Smart windshield. A windshield capable of highlighting objects and people from the outside, such as dangers to avoid or road signs, and also providing some car-related information.

Utopian ideas

- Travel back in time to create opportunities to spend more time with loved ones.
- Home–work teleoperation system.
- Extending the length of the day if necessary.
- Preventing aging so everyone has a chance to work to the end.
- Omnidirectional wheel.
- Roads able to capture the potential energy on its surface and harvest it effectively.
- Wireless power transfer by drones. The charging problem can be a big issue. The solution is for the cars that stay on long roads because the battery is dead. Meanwhile, any oil drone technologies are expected to come into play. Whether the drones are in motion or stationary, this problem can be solved by hovering over the charging point of the vehicle and transferring the power.

High-budget ideas

- Creating specific areas of the city designed for people with disabilities where they can feel safe and independent.
- Remove all manual cars instantly.
- Intelligent system of automatic telephone calls based on emergency.
- Automatic parking through sensors able to evaluate the available space.
- Braking aid through precise sensors.
- Automated cars with wheelchair access, both for the driver and for the passenger.
- Glasses with machine learning characteristics for the visually/auditive impaired – smart windscreen
- Machine learning prosthetic.
- Drone service for disabled.
- Providing autonomous vehicle service that can meet the needs of disabled and elderly people for settled communities that cannot access autonomous vehicles.
- Autonomous life-saving system. Being alone against diseases that may occur instantly is the fearful dream of every person. In this regard, autonomous vehicles should establish trust in people. For example, sudden heart attack. There are sensors in the vehicle that can measure the user's body temperature and heart rhythm, and, in case of a sudden illness, it would take him to the nearest hospital. In addition, thanks to the IoTs, there is a system that stores the health history of the user in the cloud system in the database. Thanks to the user's health history, which is shared privately with the hospital where the user is taken, it can save lives faster.
- Waterproof electronic system inside autonomous vehicle. This idea is for the users who want to eat and drink in the vehicle. In case of sudden braking or unexpected

vehicle movements on the road, the liquid may spill and electronic devices might be damaged, which can also cause problems while driving. Solutions that can prevent this destruction can be developed by taking advantage of the flexible electronics that is not affected by liquids and thus protect electronic products.

9.5   Prototyping and testing

In automotive industry, prototyping and testing play the crucial role. While developing new ideas, designers and engineers always want to visualize their concept. The whole industry believes and implements Chinese proverb "One picture is worth 1000 words" on daily basis. Therefore, the application of those two DT stages seems to be a very natural and common way how to present the idea to the wider audience, including potential users.

9.5.1   Research on automotive first prototypes

The first prototype that presented the selected concept was in the sketch form. It was the fastest and very low-cost way to show the idea in the more communicative manner. Figure 9.8 depicts chosen solution, that is, a smart windshield.

The smart windshield is a solution that allows users to have control as in usual car (typical for all people). However, it can be treated as a step towards the future – fully autonomous cars. Taking into consideration interviews from the empathy stage the smart windshield meets the requirements of the potential users. They are not ready for very advance technologies. They need more time to adopt themselves to it step by step.

The first elaborated prototype consisted of:

- a smart glass,
- a car interface system,
- a mobile phone interface system,
- and a camera system.

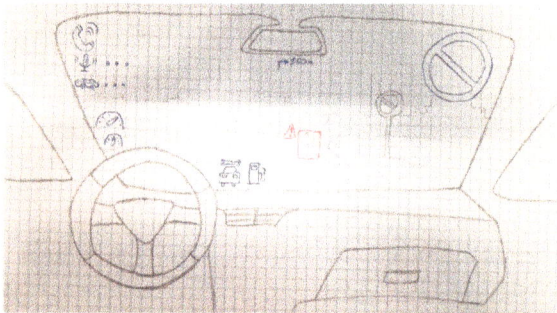

Figure 9.8   Chosen idea for further investigation and development (i.e., the first sketch of a smart windshield)

The smart windshield exploits the capabilities of an autonomous car sensing the environment, highlights dangers on the road, projects the nearest road sign more clearly, and, finally, shows some information related to car functioning. Figure 9.9 presents the design of the solution with its functionality. At the concept stage, it was assumed that there will be the possibility to customize the interface through a mobile App. The user can remove or enlarge certain texts or icons at will. Nevertheless, some limitations will exist because of safety reasons. What is more, the windshield can become darker to serve a similar role as sunglasses – the user may adjust the level of darkness according to his preferences. The whole screen with visualizations (icons and texts) will be designed with the respect to driver's field of view avoiding distraction and wide-range eye and head movements.

To illustrate the situation when the passers-by suddenly appear at the pedestrian crossing, additional variation of the smart windshield was prepared as shown in Figure 9.10.

9.5.2 Testing

Testing was treated as an inseparable part of prototyping and allowed students–designers to gather feedback about the selected concept from potential real users. The testing stage consisted of the following sub-steps:

- preparation of testing scenario: crucial here was to answer such questions as Where to conduct the testing? When to do it? How to present the smart windshield in neutral manner and to let user to have a kind of "interaction with the windshield"? Who will be the testers? Are there any requirements for tester selection? How to write down the feedback (any special tools)?;
- conduction of the testing;
- analysis of testing results. The students decided to come back to real users who were interviewed in the empathy stage. The group of taken people had the possibility to familiarize with the smart windshield, shared their opinions and asked

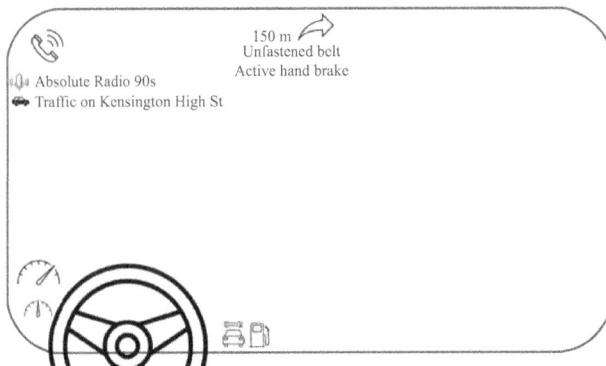

Figure 9.9 The view of the smart windshield

Figure 9.10   The smart windshield in a particular situation – a passer-by closer to the road

Figure 9.11   Four categories of a feedback-capture grid to analyze the feedback about the smart windshield concept

some questions. In order to organize the users' feedback, the very simple tool – feedback-capture grid was applied (cf. Figure 9.11).

The testing with real users revealed several important aspects that should be taken into consideration while improving the developing concept:

- users who are interested in the newest technologies are eager to try and use the smart windshield as soon as possible, they are strong supporters but still the issue of reliability raises doubts;
- more sceptic users appreciate the fact of full control on the human side – the proposed solution is a sensible middle way (people are still not ready to rely on fully autonomous vehicles);
- the majority of testers (above 90%) believes that the smart windshield may significantly increase the safety on the roads; however, there were some concerns if some elements on the screen are too distracting for the driver;

- almost all testers mentioned the cost of the smart windshield, high cost can be one of the biggest disadvantages and a huge barrier to conquer the market;
- people prefer such indirect solutions which do not change diametrically the existing world order but at the same time they accustom people to new products and way of living.

Iteration is one of the paradigm of DT. Therefore, the elaborated solution has to be improved with the continuous assistance of potential users and keep up with the newest technologies tested by many companies. After first testing session, the students–designers decided to add additional safety system that will highlight the sudden dangers on the road. The concept of adjustable transparency must be also revised.

9.6   Conclusions and reflections

DT methodology can be successfully applied in automotive industry, in the future innovation creation in the area of autonomous vehicle as well. The most distinguishable element is user- or human-oriented approach to the given challenge. Through the whole designing process, a user is the central point while other issues are spinning around and they should be adjusted to the real humans' needs.

In the frame of the presented project, students successfully followed five steps of DT process which were proceeded by the research on the current state-of-the-art in the field of autonomous driving vehicles. In such technology-oriented projects, in most cases, it is advisable to conduct preliminary research on the developed topic area, especially when there is no specialist in the field within the team.

As mentors, we observe that such DT challenges gave engineering fields' students of a new perspective on how to use their technical knowledge and hidden creativity deposits to result in several innovative solutions. A good theoretical background in engineering is necessary to follow the constantly changing world; however, empathy is something that makes the difference. Through this project, the students released that people may need some time to get familiarized with the newest technology and it is better to provide solutions that support them in adapting to the modern reality. DT methodology develops the ability to see things through someone else's eyes and to create real value for humans. It is an invaluable human-centered approach worth spreading.

Chapter 10
Design thinking in bioengineering
Dorota Bociaga[1]

Activities for the health and life of people are one of the priority areas of activity of each country. Biotechnology companies consider which courses of action are the most appropriate to respond to people's needs most accurately. The design thinking (DT) methodology is advantageous in defining these directions and specifying the real requirements of society in terms of their health and quality of everyday life.

DT combines different disciplines to find solutions within a complex and multi-layer system of business, technical and human contexts, to lead ultimately to products and services that people need, want, and are willing to adapt to their lives. In the case of products and services that improve the quality of our lives and affect our health, the faster and more readily they are applied.

People and their needs are at the very beginning of the DT process. The solution is always aimed at meeting the final beneficiaries' real needs and solving their problems. In the case of biotechnology projects, this needs to focus on people is natural and all the stronger.

As Tim Brown said, *Design Thinking is a method of meeting people's needs and desires in a technologically feasible and strategically viable way*. Methodically undertaken biotechnology projects allow for the development of commercially justified, but above all, solutions needed by people, taking them into account at every stage of the project.

This chapter presents valuable facts from the implementation of projects in the field of bioengineering, the use of which is important in the DT process and has brought measurable benefits for solutions that directly affect the quality of life and human health.

10.1 What challenges in the field of bioengineering are the most valuable for Students in the process of learning?

Students in fields such as biomedical engineering, biotechnology, and others that are oriented toward a greater or lesser extent toward the combination of medical and

[1] Łódź University of Technology [DT4u], Poland

engineering aspects are always the most willing and easy to learn the DT methodology through projects thematically strongly related to the human being, the patient, the doctor, and the medical device. Empathizing also comes more easily to them than to Students in other fields of study, as the patient and his or her welfare is a pillar of their motivation for undertaking such studies rather than others, which largely determines their ability to strongly direct their thinking toward building solutions that serve people.

However, although the subject matter in the field of bioengineering projects is relatively easy to define in an almost intuitive way, the skill to appropriately match the topic to the level of knowledge and experience of the Students is a considerable challenge for the teacher and determines how realistic and advanced the solutions will be created, but above all, how effective and built with understanding will be the process of their acquisition in terms of the knowledge in the methodology and the possibility of its later use in industry projects carried out by graduates already as employees.

In the DT methodology, it is preferable if the Team itself ultimately decides what the problem issue is. The area in which improvements are sought should be outlined broadly enough so that Students can demonstrate the ability (most essential in the DT process) to diagnose problems on their own and the ability to apply tools to identify the priority concern. In the case where the Team decides for itself what it is going to work on and is involved from the very beginning in the process of data collection and the emergence of the core topic, there is clearly greater commitment and determination visible at every subsequent stage of the project. This does not mean at all that the team does not make mistakes or go in the wrong direction – this also happens. It is then the role of the leading methodologist to, in due course, from an "outsider's" perspective, help the Team see that other dependencies also exist that are perhaps more relevant to the case being considered. The teacher's role is also to keep the Team from wandering down the wrong path for too long because the danger is that the Team will exhaust its energy reserves for wandering and will approach the next steps with less commitment or even reluctance. Balancing the time to make mistakes is important – it is not time wasted. It is time to collect data, learn and draw conclusions.

10.1.1   *Subject areas of the projects*

What subject areas of projects to choose, and how to choose them? Projects may have their source of origin in two kinds:

1. Projects, the subject matter of which is defined by the teacher – based on his or her conducted research projects, ongoing cooperation with other entities, etc.
2. Projects for which issues come from "outside" – from the implant manufacturer, doctor, health unit, etc.

First of all, before making an indication, whether it will be a topic related to diabetes, hearing defects, problems in urology, burns, or implants for cardiac surgery, it is worth reviewing what resources we have available in the aspect:

- possibilities to reach relevant locations from the point of view of the project, where Students will be able to carry out the observation process if it is not a matter of public cases;
- opportunities to contact and interview representatives of stakeholder groups who will be involved in the problem (opportunities to reach patients, nurses, and doctors at their place of work and outside);
- opportunities for immersive engagement in the project or contact with as many people as possible who can serve as "experts" and those directly affected by the problem at hand.

10.1.2   Criteria for selection of project topics

The first criterion for a good selection of a project topic is **availability**. This means that in the subject area which will be specified, Students will have the opportunity to conduct interviews freely, will be provided with access to current data, contact experts, there will be an opportunity for them to observe both patients and doctors in the course of conducting diagnostics, performing procedures, and during the recovery period. If an issue is selected within the scope of which the Team members will not have the knowledge, experience, access to data will be limited/obstructed, and in addition, they will not be provided with the opportunity to empathize through direct contact with groups that play important roles in the problem issue, their work will be difficult, based only on data acquired indirectly and may be discouraging. A high level of engagement on the teacher's part can help the project to be completed successfully. Nevertheless, the Team members' experience brought as a result of the project will result in less experiential learning.

The second criterion for determining that a topic is correctly selected is **adequacy level**. A methodist will work in a different way with a group in the first year of studies and in a different way with Students completing their studies. For the first ones, teaching activities will have to be separated into those that focus on teamwork skills, communication, task management, resources, goal setting, and milestones. Thus, substantive activities within the scope of the project's subject matter will be shared, which means that it will not be possible to devote as much time to them (this aspect can be taken into account by increasing the number of hours for the project). Working with people who have already completed several projects and are experienced will ensure that the focus can shift to the substantive aspects while smoothly introducing the methodological steps. Therefore, it is necessary to skillfully adapt the project topic to the level of preparation of the Team members that implement it. Thus, selection, or allowing the group to decide to deal with a problem that is too complicated, will most likely not allow them to come up with a solution that will be satisfactory to them, and the teacher will not be able to introduce a solid foundation in the field of project work.

The third criterion is **team competency**. It is known that the most efficiently working Teams are those that are sized according to the level of complexity of the problem and that have complementary competencies (in the persons of these Team members) from engineering, medical, IT, and other fields. Suppose the topic of the project, for example, derives from the development of biodegradable polymers, and we have only medical Students on the Team. In that case, it will be difficult for them to make a good understanding of the project topic and difficult to find their way around the development of a solution. Therefore, when choosing a project topic, it is worth taking into account the criterion of knowledge and experience that the members of the Team have and determine it in terms of their capabilities; alternatively, if there is a possibility, it is worth inviting other members with different competencies to the Teams and provide them with contacts with experts in the topic.

10.2   The process of empathy in bioengineering issues projects

In projects where we deal with studies for the direct improvement of the condition, health, and comfort of people's lives, the empathy stage is even of greater importance regarding the developed solution's relevance. It is also, of course, of great importance to the process of systematic project implementation itself. At the same time, it is a moment of verification of knowledge in the fields of medicine, engineering, and related fields and is an excellent way for Team members to learn involuntarily new facts and expand their scope of competence. Therefore, if the project is implemented as a didactic process, it will make for a very dynamic and effective learning process.

Reaching the stage of developing a solution is a moment anticipated by the group, to which they involuntarily jump from the very beginning of their consideration of the topic. This is among other reasons because (seemingly, however) this stage is easier for them. Being unprepared for it (without going the whole way of diagnosing the problem), they only rely on solutions encountered so far and do not relate this to the actual problem. It is necessary to accept this calmly yet firmly interrupt the process of thinking about the problem through the prism of proposals scrolling in the head and spoken (dropped) out loud, like "I know how we will do it – I saw such a thing recently...," "We can do it like this..." and get the Team back on track of thinking in terms of the problems that people face in a given issue, and which would need be addressed.

Let us consider the preparation and execution of the empathy stage and the formulation of problem issues based on an example of a specific project (case study) in the field of bioengineering.

As part of a class at Lodz University of Technology (Poland), the Biomedical Engineering course students were given a problem issue formulated as "Artificial kidney – dialysis." The topic was formulated in such a way that it represents a sufficiently broad topic for Students to significantly expand their knowledge in the fields of human anatomy, artificial organs, supporting devices, the role and function of the kidneys, treatment methods, and modalities (including transplantation), and everything related to this issue, also including material solutions and devices. At the same time, the topic was very well prepared by the teachers (from the substantive and logistical side) so that the Students' freedom of action and knowledge acquisition took place within the safe framework of their competence and level of preparation for this project.

10.2.1   Preparation for the empathy process – the role and possible range of activities of the teacher

In business projects, the success of their implementation in terms of the developed solution will be constituted, among other things, by the appropriate selection of Team members – the complementarity of their competencies in the field of the subject matter of the problem being solved.

In teaching projects, the opportunities for selecting people for the project are significantly limited, if they exist at all. However, the success of the project's implementation is not so much determined by the functionality and ability to implement the developed solution but by the educational effect, i.e., the amount of knowledge and experience acquired by the Students and their skills to present this in the form of a prepared solution proposal. Therefore, the role of the teacher is even more important, and his or her actions in a significant way can compensate for the lack of diversification of Team members' competencies and contribute to the quality of the didactic process and the solutions developed.

Before the topic "Artificial kidney – dialysis patients" was presented to the Students, the teachers prepared themselves to lead the Teams to work on it, as follows:

1. they verified topics related to the issue of kidney disease,
2. they identified stakeholder groups and potential beneficiaries,
3. they analyzed various scenarios estimating in which directions the Teams may develop the project,
4. they analyzed the study program completed so far by the Students to verify what knowledge they should already demonstrate and what knowledge they may be lacking (i.e., which they need to be supported with).

Considering the data collected as part of the activities mentioned above, the teachers performed the following preparatory procedures:

- they agreed on lectures by invited experts – doctors who work daily with people with kidney failure/defects, who perform transplants of this organ, engineers from companies that manufacture dialyzers and other support devices,
- time buffers and finances were secured for possible lectures and meetings with external experts from other areas who may prove crucial during project implementation,
- working hours were secured in the project for consultations with other lecturers whose expertise might be needed at separate stages of project implementation,
- specific locations were arranged where Students would be able to access in order to conduct observations of dialysis processes, to listen to conversations between the doctor and the patient during the diagnostic stage but also in cases of patients' follow-up after organ transplants (clinics, hospitals, dialysis stations, etc.),
- secured locations and time for meetings to carry out interviews with various stakeholder groups – patients themselves, doctors, nurses, technical people at the hospital, and equipment suppliers,
- bearing in mind that kidney disease can affect both adults and young children, care was taken to prepare appropriate consents and locations for conducting interviews

in a safe and comfortable manner for patients (if young children, then under the supervision and presence of a caregiver) coordinating the time of classes and the attendance of Students with the time of staff and the functioning of the institution,

- it was agreed with the experts and the groups that will be interviewed, their availability during the testing phase and the presentation of solutions in order to get constructive feedback,
- contracts were signed with companies, and time availability was arranged with their sales representatives or people from the product and research and development departments so that they could make presentations on the solutions they offer and the scope of development directions,
- they performed a current market analysis in terms of available and functioning solutions in order to be able to verify the validity of the Students' considerations and possibly lead them out of their erroneous thinking.

Project implementation requires a lot of work from Students, which mainly results from the need to search for information, acquire and assimilate new knowledge, analyze and compare data, and apply new methodological tools introduced or recommended by the teacher.

For an academic teacher, systematic management of projects in the field of bioengineering requires equal (or maybe even more) effort and time. A well-conducted project process runs in such a way that the Student performs one work after another, in which he or she uses the tools introduced by the teacher, feels that the project is developing, gaining pace and momentum, and almost does not feel how smoothly and naturally moves to the following stages of the design thinking process. Such a perception of the process is a credit to the teacher and means that he did a tremendous amount of excellent and necessary work before the project topic was even presented to the Student.

10.2.2   Team building – what to take care of and how to make a selection of members

Students in technical fields sometimes have trouble engaging in the process of empathy. They use the argument that they are not psychologists, and creating empathy maps based on interviews and drawing conclusions from them in order to do a secondary synthesis (which aims to identify the core problem) is quite a challenge for them. Meanwhile, students with less scientific minds, high openness to interpersonal relations, and high communication skills perform well in the activities assigned to the empathy stages (observation, interviewing, immersion). That's why Teams composed of people with different competencies and diverse personality traits work very efficiently because they complement each other in project work by taking on the most natural functions and roles for each other.

It should be noted that in the didactic process, the selection of people for the project is very limited. Students from the same specialty/field of study will be fairly homogeneous in terms of the range of knowledge they have, but they will certainly not be identical in terms of their personality traits and natural predispositions regarding group roles. Therefore, if it is not possible to select people with a variety of competencies and knowledge for the project, it should be taken into account that in this regard, the Team will need greater support from the teacher, a longer stage of knowledge acquisition and the opportunity for more frequent and numerous contacts with experts in the field of the project topic. However, what can and even more so should be cared for in the case of a group that is homogeneous in knowledge is to take into account the diversity of its members in terms of personality traits and to skillfully distribute people into groups, making sure that each group receives for its composition people who are different in terms of the roles they naturally play in Teams (natural leader, man of action, practical organizer, man of ideas, man of contacts, judge, man of group, perfectionist).

The teacher must assign people to Teams at the very beginning of the work, which means that he or she does not have the possibility to get to know the Students and therefore does not have an understanding of who in what role will perform the best.

How, then, can Teams be built while ensuring the correctness of the process of their work, the comfort of its members, and/or possibly, also building new competencies in Students in terms of their Team roles?

This can be done in at least two ways, taking into account whether we are dealing with upper-year Students (who already know each other well) or a freshly formed group of people who have never worked together before.

10.2.2.1   Building Teams from people who know each other and have worked together before

In this case, assigning people (who know each other) to Teams by the teacher (who does not know these people) will be acting "by force." People who know each other generally know to what extent they can rely on whom. They have already "experienced" each other in cooperation and do not want to "learn" other people or even know with whom they are not able to work. Assigning people to a group at that time will make them irritated, and instead of focusing on project work, they will waste a considerable amount of energy on building their position in the Team, or they will take a passive role, which will weaken the Team and contribute to conflicts. In such a situation, it is better to let the Students independently select themselves into Teams. Thanks to this, one skips the selection stage and can go straight to work knowing that there is a high probability that the people who have matched up and constitute the Teams already have developed channels of communication and maybe even their own language of communication, which will only speed up the project. However, such a situation does not change the fact that the assigned topic to a given Team may not quite match the level of interest of all its members in equal measure. This can be dealt with by leading to a situation where each person individually decides what topic he or she wants to deal with and, in deciding this, is fully aware of what Team he or she then joins.

What might this look like in practice?

1.   We ask a group of Students to divide themselves into Teams of four – max. six members. Some groups (with slightly shorter lengths of shared experience) need

a few days to form. It is worth giving them this time without requiring them to do it "on the fly" and under time pressure.

2. We present the Project topic to the Teams – let us go back to our case study – that is, for example, the broad topic of "Artificial kidney – dialysis patients." All Students in all groups have a similar level of knowledge and experience resulting from their course of study, which means that they start the project from the same level of substantive preparation.

3. All Teams begin their work by identifying as many kidney disease-related variables as possible by completing the preparation stage for undertaking the project (Section 12.2.3). Of course, each Team will develop this stage slightly different way. Nevertheless, everyone generally realizes the same thing reaching a level of substantive preparation, allowing them to move to the next stage or the beginning of empathy.

4. After performing extensive empathy and based on the data collected from the topic understanding, the Teams must specify the problems they have identified and which they consider the most important. Depending on the topic and intensity of the Team's work, there may be more than one problem issue. Then the Team can be asked to choose the two that, in their opinion, are the most relevant. If the Team sees a lot of minor problems, the teacher can help them connect them together.

5. The next stage I call the "stock market of problems." Teams are asked to briefly present to members of all other Teams what problem(s) they have specified and write them on the board in the form of one sentence capturing their essence (cf. Figure 10.1). The presentation should be short (3–5 min), and presenters should focus on the two most important aspects in it:
    (a) what problem do we see,
    (b) justification of why this problem is important.

6. Once the problems of all the Teams get on the board, voting takes place. Each person has three votes they can distribute freely (all for one problem, one for each of three different ones, etc.). Voting is done openly by marking one's votes next to the problem written on the board. If there are many Teams and many issues on the board, the number of votes each person will have at his or her disposal when voting can be increased accordingly.

7. After the votes are completed, they are counted, and the problems are ranked – those with the highest number of votes go to the top spots of the list. Depending on the number of people we have and who will be divided into Teams, the number of problems from the top of the list goes into consideration and their assignment to Teams.

8. The final stage involves the final selection of the project for implementation. This is the moment when each person may change the Team and decide what project (of which the topic to be solved will be the chosen problem issue) he or she wants to get involved in, keeping in mind the subject matter that is closest to him or her. Here, Teams previously formed by Students may cease to exist and completely new ones may be built. It is at this stage that individuals sometimes have to make a decision, in which they answer the question, "do I choose more with whom I will

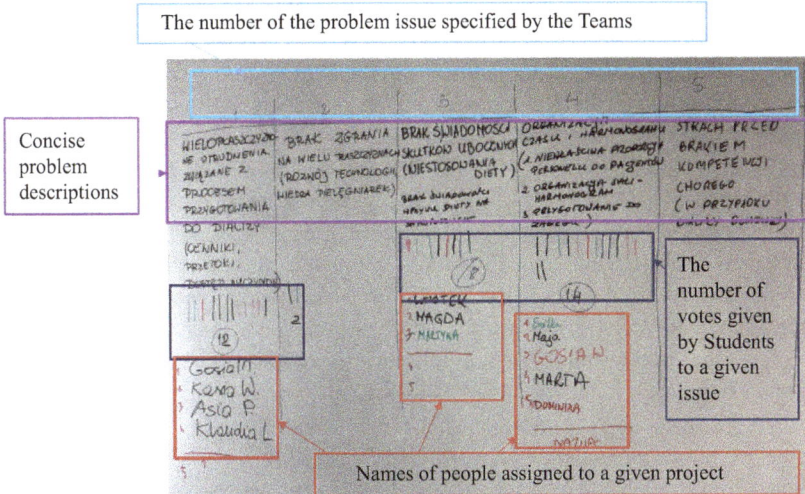

Figure 10.1   An example of a "stock market of problems" created by the Students for the topic "Artificial kidney – dialysis patients" with the results of the vote marked and the people assigned to the project.

carry out the project or rather a topic that interests me more." The choice is made based on the current presentation of all data in an open manner. It is often a difficult choice, but it always remains a choice and not an issue imposed by the teacher.

In my more than 10 years of practice working with Students during the implementation of projects with different methodologies, I have observed that their involvement is always greater, and thus the work is more effective when they decide for themselves make the decision both as to the membership of the Team and the topic they are implementing. Even if either of these two elements is not the choice that is the most optimal for them, because of the fact that they make it themselves, they take full responsibility for the choice they make, which in practice means that even in crisis situations, the effect of self-discipline and self-motivation works due to the fact that it was their own decision and not something imposed by the teacher or someone from outside. Of course, anyone can make a mistake in their choice, and in projects, difficult situations come along against which even the best Team lineup can experience a crisis. The role of the teacher is to come to the help of such a Team. However, a sense of responsibility and awareness of their own choice, which are in the Team members, makes them overcome the crisis faster. Therefore, skillful leadership of the Teams consists largely of consciously observing their autonomous actions and providing assistance when it is needed and does not disrupt the process of cooperation.

10.2.2.2   Building Teams from people who are just getting to know each other and have never worked together before

When a project is to be carried out by people who have never worked together before, it is difficult for them to decide with whom to form a Team. Their choice is random and not based on any practical considerations. In such a situation, assisting students and making a top-down division into groups is worthwhile. What to follow if the teacher has also never worked with these people before?

In such a situation, it is worth spending the first few hours of class doing some communication and teamwork exercises with the group to get to know the Students better, discover their strengths, and let them get to know each other. If there is an opportunity to use the support of a psychologist, it is worthwhile to involve one in the process of determining the natural aptitude of each person. If such help from a professional is not possible, it is a good idea to support oneself with publicly available psychological tests and open-access platforms, where Students can take a personality test or tests determining group roles. Based on their results, Teams can be compiled, taking care that, for example, natural leaders, perfectionists, judges, etc., do not all end up in one group but rather that this distribution is as even as possible between groups.

In each of the above cases (cf. Sections 10.2.2.1 and 10.2.2.2), it is worthwhile to introduce rotating role changes when projects are implemented as part of the teaching process. Depending on how many weeks (meetings), the project will be carried out, the Team will choose from among themselves a leader and a person who will be responsible for taking notes during Team meetings. Every 2–3 weeks, there is a change of roles so that each person can prove himself or herself at least once as a leader, who directs the entire Team, its work, makes decisions, and distributes tasks. This approach gives the Students experience in the form of finding themselves in different roles in the Teams where they will come to work in the future in their professional path.

10.2.3 Preparation and execution of the empathy process – best practices for Students

10.2.3.1 Substantive understanding

If the members of the Project Implementation Team are not specialists in the designated subject area of the project, it is worthwhile for the empathy stage to begin with extensive substantive understanding (cf. Figure 10.2). Available market solutions, the latest scientific developments, the most common problems, their scale, system support, the situation in different countries, and any statistical data that will help to assess what and to what extent are novelties and challenges in a given topic. A good overview of the topic will pay off at each subsequent stage of the project and will speed up its implementation. Omission of this point does not preclude the possibility of undertaking the project, but deficiencies in understanding will hinder and slow down the work of the Team after a certain level of task completion. Sooner or later, a well-organized Team actively implementing a project comes to a point where it finds it is necessary to do research on the topic. The research carried out prior to the tasks in the field of preliminary identification of problematic issues related to the topic of the project, stakeholder groups, and activities such as observations or conducting interviews incomparably accelerates the work in the project and represents the professionalism of the Team in contacting people from outside who will be involved in the implementation of the empathy stage.

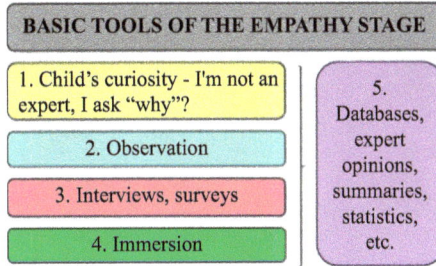

BASIC TOOLS OF THE EMPATHY STAGE

1. Child's curiosity - I'm not an expert, I ask "why"?

2. Observation

3. Interviews, surveys

4. Immersion

5. Databases, expert opinions, summaries, statistics, etc.

Figure 10.2   The basic tools that can be used in the empathy process to best determine the needs of the user and thus diagnose the problems they face, the solution to which will represent the scope of the project

Let us refer to the case example from the topic "Artificial kidney – dialysis patients." In the first step of the implementation of the topic analysis process, Teams began to determine with which the issue and related problems are connected. This step very quickly and clearly outlined the degree of orientation of the Team members on the subject of kidney diseases and their treatment methods, both in terms of knowledge and personal experience. This gave the students an initial idea of which areas they must analyze the data and gave the teacher an indication of which preparatory steps (described in Section 12.2.1) should be started and to what extent to implement them.

10.2.3.2   Identification of stakeholder groups

After the preliminary identification stage of the topic related to kidney dysfunctions and dialysis processes, the Teams identified stakeholder groups that would be relevant to be able to identify and clarify the problem issues (cf. Figure 10.3). These groups were involved in the empathy stage. The order of who will be interviewed first was determined based on a stakeholder map that considers the relevance of the factor of interest in the problem and the degree of influence by the identified groups.

Because the Teams considered both patients with congenital and acquired defects, no age group was excluded, and thus interviews were scheduled with both children and adults. In addition, for meetings and interviews, experts from the medical side (doctors, nurses, dialysis station operators, etc.) and the engineering and logistics side (dialyzer manufacturers, people who direct the movement of donor organs) were identified. The students met with diabetes patients and their caregivers. They came to the hospital and dialysis stations for observations. They met with experts in the field of dialyzer construction, directors of the facilities that decide on the organization of dialysis, and the dialysis space itself.

In projects where the scope of the problem is narrowly defined (specified) from the very beginning, stakeholder groups can be narrowed down right away. This will involve either short teaching or business projects contracted by a particular business/institutional entity. As a rule, however, we then have only one Team that deals with it, and its goal is to obtain a solution to the problem issue as soon as possible.

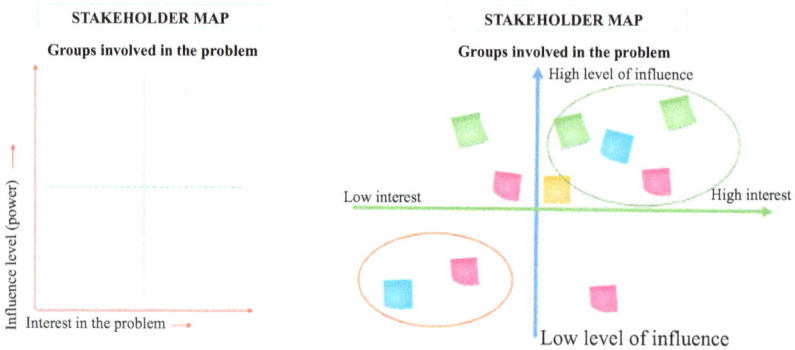

Figure 10.3   Stakeholder maps to be filled in at the stage of understanding what groups are involved in the problem issue and what their impact is on it (on the left). Filled stakeholder maps with highlighted groups (top left corner) that have the highest impact on the problem and are most interested in eliminating it.

The Team will be composed of experts adequately selected for the subject matter of the problem. In projects with a didactic background, where the goal is to acquire knowledge, the longer path of arriving at the problem issue (through understanding the topic at the empathy stage) and an extended path in terms of developing a solution are the most beneficial from the point of view of learning, acquiring competence and experience. And thus, the ways of carrying out the above-described stages of the design thinking process can (and even should) be diversified depending on the assumed purpose of implementing the project with this methodology.

10.2.3.3   Preparation for observations, interviews, and surveys

"User" in the empathy stage – technical aspects.

Before the interviews began, the Teams met with a dialysis equipment manufacturer representative. They listened to a lecture by a representative and then met with a representative at the hospital (where artificial kidneys support patients before surgery), at the dialysis station (where patients come several times a week), and were able to familiarize themselves with dialysis devices of different generations, their various components and learn about the principles of operation. This stage gave the students the strength to analyze the technical aspects of "artificial kidney" type solutions, that is, to expand their knowledge in the field of engineering sciences. The technical basics outlined new aspects of the issue to the students, which they added to their lists of issues related to problems associated with the dialysis process.

"User" in the empathy stage – biomedical aspects.

In the next step, the Teams moved on to the empathy stage associated with observations and interviews with groups directly involved in kidney disease, transplantation, and dialysis processes (cf. Figures 10.4 and 10.5). Before proceeding

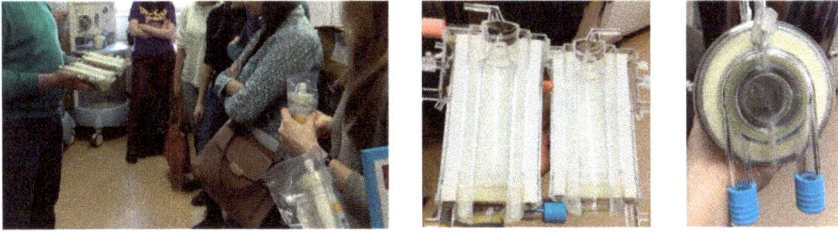

Figure 10.4   Students learn about the technical aspects related to the construction of artificial kidneys (presentation by an expert in the field of technical solutions), their individual components (photo on the right), how they are operated, and their integration with the hospital's central systems (as a factor affecting working conditions and opportunities for a change)

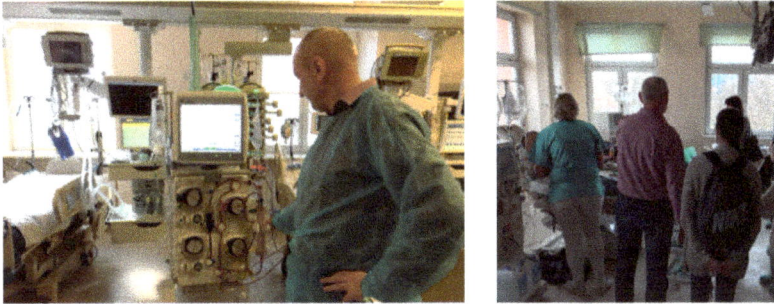

Figure 10.5   Students learn about the types of dialyzers used in hospitals (photo on the left) and dialysis stations (photo to the right), how they are used, how they are operated by medical personnel, and the technical possibilities they offer

with interviews with each of the identified groups (a child with kidney defects, a parent/caregiver of such a child, a doctor performing transplants, a nurse connecting a patient to an artificial kidney, etc.), the Teams prepared above all in terms of well-chosen and formulated questions, making sure (under the teacher's guidance) that they were appropriate to the type of information they wanted to obtain and/or verify, but also as to the distribution of tasks – who would conduct the interview and who would be responsible for taking notes, who would prepare consent forms for taking photos, recordings, who would collect them, who would take and archive the collected materials (cf. Figure 10.6). The teacher's main task, in turn, was to prepare an admission to the designated locations and manage the timing of visits, interviews, and meetings, as well as to indicate rules of behavior, such as in terms of the prohibition of asking sensitive questions, maintaining complete privacy for those being observed, and in terms of hygiene rules in the locations where they will go.

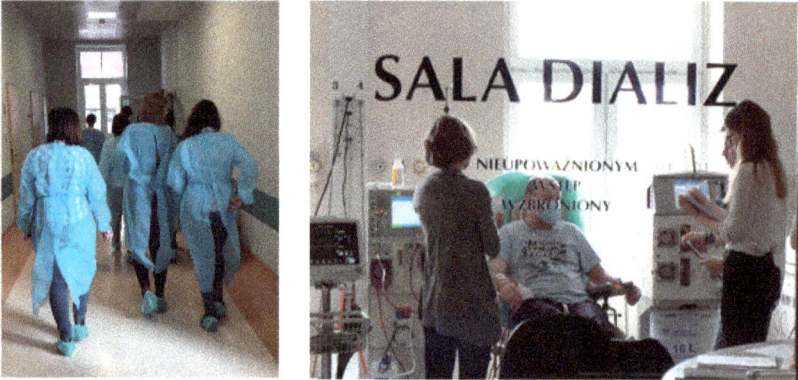

Figure 10.6   Students heading out for observations and interviews with doctors and patients in the hospital (on the left) and a teacher interviewing a kidney transplant patient who has been on dialysis for many years (photo on the right – sign on the door "Dialysis room")

10.3   Identification of the core problem

The empathy stage can be very time-consuming. The time needed to analyze the facts collected in this part of the project ends up with the formulation of the core problem issue in the area of the topic given at the very beginning, i.e., the determination of the so-called PoV (point of view), which in the design thinking process is separated, as the next stage. De facto, this is the point of contact at the edge of the empathy stage that is ending at the moment and the stage of developing a solution to the main problem that is about to take place. A thoroughly carried out empathy stage allows for identifying the problem accurately. Because it is a stage of gaining knowledge, data analysis, and finding connections for non-obvious facts, it very often takes as much time (or even more) as the other stages of the DT process. Nevertheless, it is worthwhile to devote enough time to it, as it carries many cognitive aspects and in the knowledge gained through contacts with experts, ideas are also involuntarily built that will be applied during the development of the solution. Albert Einstein himself once said, "If I had an hour to solve a problem, I would spend 55 minutes thinking about the problem and five minutes thinking about solutions." His point is an important one: preparation is of great value in solving problems. Therefore, it is worth spending time on the empathy stage, which is a kind of preparation for further actions.

How to conclude the results of the preliminary and the secondary analysis in the empathization process to obtain POV was described in previous chapters of this book. However, in the aspect of bioengineering projects, it is worth paying particular attention to identifying correctly: for whom are we designing? For whom are we looking for a solution? Who is supposed to be the end user?

It is not always the direct entity for whom the solution will be developed, that will be the patient or the doctor. On the contrary, as a rule, it will be an entity that is able

Figure 10.7   Students working on empathy maps in a secondary synthesis process that aims to specify the most relevant problem (photo on the left) and students verifying and refining the emerged problem through the "5 Why's" tool (photo on the right)

to implement a solution to introduce a specific facility, a product, or an application from the technical side. Therefore, when designing them, it is necessary to consider the entity's capabilities, resources, and goals. Nevertheless, the end user who will benefit from the solutions introduced by this entity will already be a dialysis patient, a person waiting for a kidney from a donor, etc.

What is also very important – often, the patient's ability to feel the change is determined by whether the people in his or her environment have been included in the design process and whether system limitations have been taken into account. Neglecting these aspects can make the solution very good but unattainable for the patient.

Let us consider this using the example of the problems and studies students formulated as a result of their projects on the "Artificial kidney – dialysis patients" issue.

As a result of the wide-ranging empathy process (cf. Figure 10.7), the Students identified some very relevant problem issues. During the discussion, keeping in mind the aspects highlighted by all stakeholder groups, they selected the issues they recognized as the most urgent to be resolved and would bring breakthrough changes for the beneficiaries:

1. Multifaceted difficulties associated with the process of preparing for the connection of dialysis equipment (catheters, fistulas, vascular access).
2. Lack of commensurate alignment at the level of nurses' knowledge and technology development.
3. Lack of patient awareness in terms of the side effects of not following a diet (the impact of diet on well-being and dialysis frequency).
4. Organizational problems in the dialysis room in terms of dialysis schedule and patient preparation for the procedure.
5. The fear of the patient's lack of competence (in the case of home dialysis).

Of the problem issues identified above, the Teams decided to implement those under numbers 1, 2, and 4.

In the case of problem No. 1, the Team has developed a new generation of highly advanced insertion system that minimizes the phenomenon of fistula formation and pain in patients. This solution simultaneously offset nurses' problems with patients' decreasing vascular accesses. It was a solution in design and materials, of which the manufacturers of such medical devices were the direct beneficiaries. Only after they implement such a solution on the market, the beneficiary becomes the dialysis patient and indirectly also the nurse connecting the patient to the device.

For issue No. 2, the solution became a simplified and more intuitive dialyzer control panel, thanks to which it was possible to make manual settings allowing the device's parameters to be adjusted to the individual needs of each of the patients who are connected to the dialyzer in turn (one after another). Newer and newer generations of devices have set more and more parameters automatically. This caused the device to pop up alerts, pause the process and call the operator when a flow problem arose (when unexpected situations arose in a patient). The introduction of simplifications is again the manufacturer's scope. However, in this case (after implementation), the nurse will directly benefit from the one introduced, who can more quickly and easily deal with problems with the device and thus can smoothly handle a larger number a larger number of patients simultaneously without generating delays in the queues of dialysis patients. This also benefits the patient, who does not feel the stress associated with alerts from the device, and his or her time spent on the bed at the dialyzer (several hours) is not extended. The indirect beneficiary of this solution is also the dialysis station itself, which can perform dialysis on time (at the scheduled time), as well as the patients' closest relatives, who, when arriving to pick up their loved ones, often have to wait several or tens of minutes longer to pick up the patient, due to prolonged dialysis. Thus, the group of beneficiaries that benefit from mitigating this problem is really long.

The situation was different for the solution to problem issue No. 4. This is because it turned out that patients suffer because of the cold during dialysis. Blood moved outside the system, plus stillness on the bed during dialysis, cause the body to get very cold. Additional blankets do not keep them warm enough, and, in turn, very high temperatures cannot be established in the dialysis room, as this is a huge inconvenience for the equipment operators and patients (i.e., people who have to move, lift, etc.). In response to this demand, analyzing all the variables and limitations discovered during the empathy stage, the Team developed special heating mats adapted to dialysis beds, with filling adapted to the varying weight of patients and an outer coating that allows their surfaces to be easily, quickly, and repeatedly sterilized. This solution could go to a variety of entities – it could be an extension of a mattress/heating mat manufacturer's product line and/or be an improved product of a medical equipment manufacturer. The people who will directly benefit from implementing this solution will be patients undergoing dialysis who suffer from body cooling during the blood purification procedure. Nurses, doctors, dialysis stations, or relatives of people with diseases will only be indirect beneficiaries of the effect of their comfort.

When the Team is in doubt about who they are designing for directly, who they are designing for indirectly, and the needs of which groups of beneficiaries they should consider on an equal level, the example of a baby carriage (or wheelchair) can be

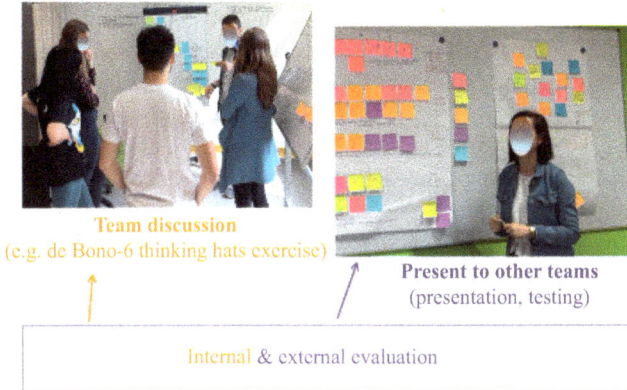

Team discussion
(e.g. de Bono-6 thinking hats exercise)

Present to other teams
(presentation, testing)

Internal & external evaluation

Figure 10.8   The stage prior to building and testing prototypes – internal and external evaluation (solution refinement stage)

cited. Such a vehicle, basically designed for a child, will not go without the drive provided by mom, dad, or grandma. Over a long distance, uncomfortable for the person pushing will become used only for short distances. So, even the best solutions used in the baby carriage for the comfort of the child for long walks and maybe even naps will not make any difference because overlooking the convenience of the other user will be a barrier to the use of this product in general.

10.4   Prototyping and testing in bioengineering

10.4.1   Evaluation stage

Before moving on to building three-dimensional prototypes (even low-resolution ones) and testing them in the broader group, it is worthwhile to carry out even earlier the stage of verifying solutions through their internal (within the Team working on the solution) and external evaluation (among all Teams working on projects from the originally defined subject area).

The Team carries out the internal evaluation for its solution in its own group. The idea behind this evaluation is to take a critical approach to its development in order to identify any errors and deficiencies, refine details and develop important functionalities. Often the Team finds it very difficult to criticize its own "child," so it is worth using some tool here, such as de Bono's 6 thinking hats.

External evaluation is still conducted in a fairly narrow circle. Teams (after refining their solution based on the weaknesses identified in the internal evaluation stage) prepare a 3-min presentation in which they indicate what problem issue they are solving, the idea behind their solution, and how they intend to prototype it. This is an essential stage from the point of view of working on solutions (it allows them to be refined even before the first prototype is created), but it is also a stage for all Students, thanks to which they continuously expand their knowledge by sharing it on the forum. Because they all work on projects within the same topic, they have a lot to add and often ask very relevant and valuable questions, thanks to which the projects are refined. In addition, they help each other develop an idea for a simple prototype, but one that most closely reflects the idea of how the solution works.

At this stage of evaluation, it is worthwhile to introduce to Students the principles of giving constructive feedback (e.g., according to the sandwich rule, "What if?" etc.). This is a crucial aspect – only properly structured feedback allows all Teams to a real improvement of solutions. In addition, such a skill (already practiced in real-world conditions) is a valuable, helpful competency in professional life.

10.4.2   Prototyping and testing stage

Giving the solution (refined through stages of internal and external evaluation) a real "shape" is worth beginning by making prototypes with the lowest resolution and then testing it as widely as possible on a properly selected group of potential beneficiaries. For bioengineering projects, prototyping and testing are governed by the same rules as solutions in other areas. What is worth noting here is the diversification of the degree of detailing of the prototype from a given perspective, depending on the group with which the testing process will be conducted. The design and material solution for the new dialysis access presented to representatives of the companies producing medical equipment will have to focus on technical aspects, availability and safety of the proposed materials, and feasibility analysis. Presentation of such a solution in front of nurses will require outlining advantages such as simplicity of use, the time needed to place such an insertion and possible complications. Therefore, the testing process will require not only refinement of the solution but also proper testing depending on the group we invite to this testing. So the three most important issues in testing are:

- think about the testing process from the empathy stage (people who give clues in determining what the problem is will also be very valuable testers who have every reason to provide constructive feedback in terms of a solution),
- ensure that the testing people represent the right group (aware and correct definition of beneficiary groups will determine the possibility of rapid and effective refinement of the solution (cf. Figure 10.8)),
- perform the tests in the right order (if the information from patients allows an even better design of the solution structure, it is worth testing in this group first, refining the solution, and only in the second step carrying it out in the engineering group).

For the next presentations (after the internal ones), it is a good idea to invite experts from outside – people who are specialists in terms of the particular topic the students are working on (nurses, doctors, engineers). Based on this, the solution is refined, and a new prototype is created: Prototype presentation. It is worth inviting experts and specialists again, and people directly interested in the solution developed. Here a discussion between invitees may (and very well, if it happens) arise. Conclusions and questions arising from it should be meticulously noted by the Students so that, based on the feedback contained therein, they can refine their solutions before the final presentation. Refining the prototype.

Final presentation (in didactic projects). It is worth inviting people to it, not only those who are already interested in the emergence of the solution but also those who may decide to implement it. So, it will be someone from the company producing dialyzers, the hospital director deciding on purchasing additional (even minor) equipment, etc.

Each stage of testing should result in a refined solution. Based on this, a higher-resolution prototype (with better materials, more detail, etc.) can be developed to move iteratively to the next round of testing.

10.5 A well-designed and well-organized learning environment is, as the third teacher

Creating comfortable conditions for interviews determines how much information can be collected. The patient's comfort translates into more sincere and more elaborate answers to questions, and this provides the opportunity to build empathy maps (cf. Figure 10.7), which allows to perform the correct conclusions at the stage of secondary synthesis and ultimately to formulate the most accurate PoV, that is, to find the real problem.

What is a comfortable environment for the Team interviewing or observing process? It is taking care of all the formalities like patients' and doctors' consents, knowing their schedules, making appointments, and well-prepared questions that accurately gather information without excessively overburdening the medical staff with the time spent on additional work for them.

Throughout the project process, the Team needs its own "exclusive" space in which it can gather all the materials it has collected. In this aspect, a fact is worth noting that many times the collected data will be subject to protection (patient data, photos, etc.), so it is necessary to prudently make this data available for possible access by third parties. The numerous electronic tools currently available reflect in a good way the mode of design work carried out in the stationary space. It is, therefore, worth remembering to work with the relevant data in a properly secured manner.

10.6 Summary – important aspects of the realization of projects in the field of bioengineering

1. An important aspect of implementing projects in the area of bioengineering is to specify the group of stakeholders and ensure that they participate as much as possible in the design process. This determines the correctness of the formulation of the actual problem.

2. Interviews in projects involving medical solutions that require patient involvement can be difficult to conduct both from a logistical and mental perspective. This is why the Team members must be adequately prepared for this and/or be supported by professionals at this stage.
3. It is worth remembering that project elaborations have direct as well as indirect users. When designing, it is necessary to take into account not only the beneficiary but also the environment in which it operates. Otherwise, there may be a situation where the direct beneficiary will not be able to use the solution, even when it best meets his or her needs.
4. For didactic purposes, in the field of bioengineering, projects can come "from outside," but the teacher can also formulate them. Here it is worth remembering that if they are real and up-to-date, it is significantly easier to get the Students interested in them, and they carry out these projects with more commitment.

Part III

Incorporating design thinking in engineering curricula

Chapter 11
Design thinking in the classroom
Íñigo Cuiñas[1]

Our engineering students should be aware that one of the consequences of living in a society is that in our daily lives, we are surrounded by challenges and problems. As the professionals who they are or they will be, they can mitigate, and even solve these problems. Thus, they can leave an imprint on society, on a small or large scale. Then, one of the missions of lecturers is to make the students aware that their professional activity is not something isolated; they could change the world with their work. Of course, there are deep and superficial changes, but many engineering proposals generate transformations in the surroundings. This effect on the society must always be accompanied of ethical responsibility.

For these reasons, creating subjects within university programs that move our students to reflect on their individual responsibility become an obligation in the mission of preparing the new generation in both technical and ethical aspects. Then, we could propose subjects that, in some way, resemble professional life and serve as learning and training our students in skills like teamwork, leadership, or problem solving. Students will acquire important competencies for their professional and personal lives, at the same time, they take care on the social and individual concerns of their works.

In addition to the benefits for students, DT can also benefit education professionals, as it offers innovative ways to plan lectures and helps to integrate technological advances in the classroom in a simpler way. According to Maureen Carroll, research director of the *Taking Design Thinking to Schools K12 Initiative* project at Stanford University [59], much of today's educational system provides training for finding the correct answers and filling in the blanks on tests, because this type of education facilitates streamlined assessments to measure success or failure. It is necessary to look for alternatives to this learning model, especially in those schools in which resources are scarce, since skills and abilities are more and more necessary to live in a changing society where problems are increasingly more complex.

Design thinking (DT) methodology results to be a good support to acquire competencies enumerated above, and that is why this chapter reflects a proposal for a course sheet or extended syllabus based on that, for its implementation in Engineering courses, at bachelor or master level. This course sheet intends to help in introducing

---

[1]atlanTTic, Universidade de Vigo [GID DESIRE], Spain

DT methodologies into a subject, like a first step bringing tools to students to be closer to working life.

This chapter describes the minimum contents of this extended syllabus, such as general description, number of European Credit Transfer System (ECTS) points allocated, mode of delivering, objectives of the course unit, skills or competencies to be acquired, learning outcomes, contents, planning, methodologies, assessment, and, finally, recommended readings.

11.1   General description

This course sheet describes the contents of a core course on DT, which could be applied as is, or integrated with a more general project-based learning course as a supporting methodology. In summary, this course sheet intends to help in introducing DT methodologies into any engineering program.

Basic structure of the course

The proposal is a three-stage course:

1. Workshop on DT fundamentals.
2. Team-working activities for developing real-world projects.
3. Assessment by pairs and lecturers in an open session.

The main framework is that the public are undergraduate students, with background at different disciplines, but not necessarily training in team working, and probably without work experience. In such situation, lecturers need to move the students to face with circumstances that they will face in real-world situations: difficulties that suffer real-world clients or users, collaboration with different groups or specialists, non-canonical or non-academic problems, and interaction with external institutions or corporations. DT arises as an established methodology that allows the explanation of these non-strictly engineering skills in a well-structured approach, which is appreciated by future engineers as it fits their way of thinking.

The objective could be that students take consciousness that the activity of the professional is not an isolated action but it transforms the world (at small and/or at large scale). Actually, the objective is to equip them with the tools to face real-world situations.

This leads to two fundamental ideas. First, the society, this means people that conform it have problems that engineers can solve. And the mission is to resolve or mitigate troubles, not to create new ones. And, second, the professional activities have direct influence in the own society, in how people live or in how they relate among themselves. This influence has to go accompanied of ethical responsibility.

With this contour conditions, the subject is structured in three parts, beginning with a workshop on DT, based on hands-on activities to learn by doing. Then, once the students are trained on DT, the proposal is to create teams of students and to address each team with an interdisciplinary project. The project is not expected to be completely defined: each team has to receive some tips on the situation and some ideas on the people that is involved. At least one faculty member will supervise each team, but in order to enrich and facilitate the cross-fertilization among different areas and methods, a number of two or three lecturers from different departments would be recommended. The lecturers guide the students along the process, following the different DT stages, and help them in preparing a test and show session with all involved lecturers and their mate students. Finally, each team will defend its work at the end of the course as a part of the evaluation process.

The assessment process is also considered along this chapter. Ideally, a procedure combining marks given by the lecturers and evaluation by peers is recommended. However, there are regulations that limit these possibilities; thus, each implementing university has to decide the best way for doing that.

11.2   ECTS credits allocation

Depending on the University where the course is implemented, the number of ECTS (a way of account for the time effort of students to pass a subject) will vary. Then, the course sheet is defined in a flexible way. It will contain a basic scheme for 2 ECTS, which is considered the minimum to provide a good training on the related skills, but it could expand till 12 ECTS by promoting the exploration of more ambitious projects.

The project-based learning focus of the topic allows the lecturers in charge to modulate the expected outcomes of students' work, and then they can ask for deeper definitions of the solutions each team provide depending on the allocated ECTS: from just a block defined system to operative prototypes tested with real-world users, passing through mock-ups, demos, or preliminary prototypes.

11.3   Mode of delivering

The course sheet is clearly defined for face-to-face delivering. The philosophy of DT needs a direct relationship among designers and possible final users, and among the team members when interact at different project development stages.

We can think that it is possible to move contents to online sessions. And, of course, it is. In fact, there are examples of online courses on DT basis, both Massive Open Online Course (MOCC) and more personalized proposals. However, a special care must be taken in this case: empathy is a must in DT developments; and face-to-face activities give an extra impulse to empathic interviews, or to group sessions for i.e. brainstorming. Online limits the body expression, the contact, and even the voice modulation details, reducing the expressivity and making more complex the team working. This is the reason of recommending the face-to-face mode of delivering.

The role of the lecturers

Lecturers in charge of a course on DT have to boost energy to the students. They are going to move students to do something, less than to just listen to a lecture.

Lecturers have to observe the activity of the students, provide feedback (more questions than answers) to them, motivate the participation of all the members of the group, and encourage them to improve their solutions.

Attitude is more important than knowledge in many occasions.

In case of being online the only implementation mode, the synchronism is highly suggested, aiming to simulate as much as possible a live communication process. This means that options in which the members of the team interact by forum applications, messages, e-mail or other asynchronous procedures must be avoided, except for some (and short) theoretical contents of the methodology. Anyway, there are some activities of the course, mainly at empathy and test stages, that must be face-to-face as much as possible (at least, it is possible to create small groups for face-to-face work, even when the activities in big group are online): it is really difficult to implement a pure online course on DT.

11.4   Objectives of the course unit

The objectives of this subject can be summarized into three main ones: basic knowledge on DT, team work, and project development. Along the following paragraphs, each of them is enunciated and explained.

Transference of basic knowledge on the DT methodology as a tool for problem solving, and for learning a systematic approach to find solutions to meet customer needs.

DT is intended as a tool for training the students in solving problems in a different way, focusing on the users and not on the engineering canonical solution. However, it is not only a tool but also a learning path that moves the students to absorb knowledge from users, environment or conditions as they advance in the definition of the problem and in the proposal of solutions. Thus, the transference of knowledge on DT methodology is more a guide for motivating the students towards a client-focused solution than an objective itself: the subject is intended to train future engineers for solving problems to people, and the methodology is a vehicular activity to reach that objective.

Transference of basic knowledge on the DT methodology as a tool for problem solving, and for learning a systematic approach to find solutions to meet customer needs.

DT is intended as a tool for training the students in solving problems in a different way, focusing on the users and not on the engineering canonical solution. However, it is not only a tool but also a learning path that moves the students to absorb knowledge from users, environment or conditions as they advance in the definition of the problem and in the proposal of solutions. Thus, the transference of knowledge on DT methodology is more a guide for motivating the students towards a client-focused solution than an objective itself: the subject is intended to train future engineers for solving problems to people, and the methodology is a vehicular activity to reach that objective.

Familiarize students with the methodology of DT based on teamwork in small groups.

We intend to train students in a practical application, not to convert them into specialists in DT methodology. Students must acquire the ability to apply different techniques, included in or related with DT, in order to work in small teams and to be able to propose solutions as a result of this team working. Students must learn how to work efficiently in teams with a clear aim: to propose solutions to a real-world problem.

The students work in interdisciplinary (if possible) group, develop a project and present the results.

Extending the public of the subject to students from different faculties would enlarge the points of view to tackle the problem and, then, the discussions along the process would be more creative. If this is not possible, the teams must combine students with different academic specialization, gender, age, cultural origin, or any other characteristic that could improve the heterogeneity of the group.

Once these teams are created, looking for variety among their members, they are committed to work together, following the different steps of DT methodology, for developing a solution to an addressed problem and, then, to explain the results to their mates and lecturers.

Each member of the group faces the problem with their previous technical background and personal skills, and improves those with the group mates'.

This objective involves a variety of soft skills that will be very useful for the students in their next future professional careers.

11.5 Skills or competencies to be acquired

The following is a list of skills or competencies that the students would potentially acquire when completing the course. Depending on the program in which the course develops, the lecturers in charge must decide the more adequate ones and select those to emphasize.

The core skills, which are suggested to be common to all the students, are:

The ability to solve problems with initiative.

The methodology itself runs away from standard or canonical solutions and seeks imaginative or different answers. For this kind of results, initiative is more than important, it is essential. This means that students involved in a process based on DT are intended to provide responses to the opened questions, and these proposals would surely require personal and collective initiative to be constructed, avoiding typical solutions.

The ability to make creative decisions.

All DT process is focused on encourage creativity: procedures are designed to move the students to imagine solutions more than to provide the solution expected by the lecturer, as it occurs in traditionally taught topics. Then, creativity is in the core of the methodology in which the students are immersed. Considering these contour conditions, creative solutions are more likely to emerge. And along the process, the different decisions the teams have to make will be more creative than those arising along a traditional project-based subject.

The ability to communicate and transmit knowledge and skills, understanding the ethical and professional responsibility of the professional activity.

This skill has different implications and requires a long comment.

Communication improves in most of the team working activities; but DT is especially interesting for supporting this skill, as it encourages a lot of personal interaction: with implied people during empathy stage; with the other members of the team during definition, ideation and prototyping stages; with possible final users during testing stage; and in the case of this subject proposal, during the defense and assessment events facing mates and lecturers. Then, improved communication and transmission of knowledge seems to be an obvious consequence of performing the training on DT and the development of the project.

The methodology itself that moves the projectors to interact with the possible final users (instead of just contact with them) during the empathy stage, encourages the students to realize the implications of their proposals within the society, taking conscience of the ethical responsibility, as they are influencing the people using their solution.

The ability to write, develop and sign projects, integrating previously acquired knowledge.

This could be considered a typical skill acquired when passing a project-based learning subject, but in this case, it has an improved significance. Students face the problem with their own weapons (their previously acquired knowledge and skills), but combining those from different people, and adding some social skills, different than those technical that they studied.

At the end of the process, students have to show their proposals: they have to write a report on the developed project, and they also have reflected on their professional responsibility, which increases their concern on the quality, the ethical, and the technical aspects of the solution. These reflections made them understand the implications of signing a project, which are longer than just doing a project.

The ability to analyze and assess the social and environmental impact of technical solutions.

Their interactions with final users and people implied in the development of solutions increases the social conscience of the students: instead of being taught on these social links, undergraduates realize that the solutions they propose have impact in the society. DT moves them to empathize with public related to the problem they try to mitigate, and to test the solution with these implied people. Thus, the social impact is in the core of the activity.

Assessment of environmental impact must be worked by the lecturers, suggesting the students to analyze their proposals from different focus, and giving them the tips to take into account the importance of checking the ecological effect of their activities. In some way, this takes part of the empathy stage: both people and environment are involved in such process.

Encourage cooperative work and skills like communication, organization, planning and acceptance of responsibility in a multilingual and interdisciplinary work environment, which promotes education for equality, peace and respect for fundamental rights.

This skill roots in the DT fundamentals. The methodology proposes a team working activity from empathy to test stages, looking for a distributed intelligence more than for a leader-guided action. Thus, these interpersonal skills (communication, team working, planning and personal responsibility within a group) have a good field to be work on them, allowing a living laboratory to experience these competencies and to learn how to manage them in a long- or medium-term group activity.

There are other characteristics of the work environment that strongly depends on the student recruitment: multilingual or interdisciplinary groups are only available in some cases, but the teaching staff can always drive the creation of groups in which members have different profiles and interests.

Equality, peace and respect for fundamental rights must be a clear objective in all team working activity. And in this case, they have to promote as part of a DT principle: empathy is a key to be involved with the final users, and it is not possible to be empathic without being respectful regarding the rest of the group members and the final users.

The ability to begin thinking about entrepreneurial ventures, from early idea to eventual implementation.

DT promotes the searching for non-canonical solutions, and alternative ideas, or new focuses, are in the origin of several successful entrepreneur initiatives. The methodology helps in the different initial steps for creating a new venture: identifying the problem of the future clients, defining new solutions, prototyping the proposal and test with these potential users are strong phases towards a would-be achievement. Evidently, only some of the students' solutions could be the first step to an entrepreneurship, but some of them can. The eventual implementation of a free enterprise would need more than the contents of the course we are proposing, but they can be a good starting point for jumping to a start-up program. Anyway, the topic can help some of the students to wake up their entrepreneur abilities, or to explore a labor world different than being employee in a company owned by others.

Additional competencies could be developed, for example:

The ability to synthesize findings and define a specific problem to be solved from a more general one.

This skill is a clear consequence of the empathy and definition stages of DT: the students have to identify the problem, and clearly define it extracting its characteristics from the contour elements.

The aptitude to manage mandatory specifications, procedures and laws.

A good topic definition does not limit the activity to a simulation of problem solving, but it has to provide the students with tools to reach a real solution. And these means that the solution has to be implementable in terms of legal issues, of fitting all mandatory specifications, and following the procedures required by the real world. Students must feel that they are involved in real-world work, not in another academic task. And that they are not joking with strange or exotic solutions but they must provide professional outcomes.

The ability to work in interdisciplinary groups in a multilanguage environment and to communicate knowledge, procedures, results and ideas, in writing and orally.

In case the conditions of the university program allow that, it could be interesting to reinforce the proposal of promoting interdisciplinary by adding a specific skill to be acquired by the students. In this case, the course sheet has to reflect this skill including a clear procedure to create the teams, forcing the presence of people from different disciplines and using different languages, in order to promote the acquisition of this ability.

The development of discussion ability about technical, economical, and/or social subjects.

The strong time dedicated to team meetings for developing the different stages of the project has additional objectives rather than the project itself: students are motivated to provide arguments on the various aspects involved. Technical proposals, with diverse focuses, will arise. But not only technical: economic issues have to be discussed, and social implications must be taken into account to reach a consensus on the proposal to be prototyped and finally tuned to be defended as final work.

The ability to elaborate the proposal of technical projects according to the specified requirements in a public competitive bidding.

Depending on the extension of the subject and the interest of local university managers, the outcome of the process could be just the presentation and defense of the prototyped proposal or going further, asking each team to prepare a complete technical project fitting all.

11.6   Learning outcomes

A collection of learning outcomes, adaptable to the program in which the course inserts is listed in the following lines.
After completion of the course, the student is able to:

Apply teamwork principles.

Taking into account the characteristics of the DT methodology, this learning outcome seems to be easy to reach by those students following the course and participating in the different proposed activities. The DT workshop itself would be probably enough to transmit the students the principles of work in a team and the ability to apply them in different situations. Besides, students are trained in team working during the development of the projects.

Actively participate in ideation sessions and generate solutions appropriate to the diagnosed problem.

The course activities move the students to participate in ideation sessions (as a training during the workshop and as an assessable element in the project development period). Thus, all students acquire experience on ideation sessions at the end of the course. Besides, they experience the process of solution generation focused on the diagnosed problem (i.e., on the results of the definition stage), concretely on the problem they must solve or mitigate with the final proposed project. The step forward from "knowing how to do it" to "doing it actively" occurs if lecturers are involved in the groups and do not let anyone be passive. Then, the full acquisition of this learning outcome depends both on the students and on the lecturers.

Prepare and conduct presentations at various stages of the project.

Along the project development, students are asked to conduct different presentations, not only the last one, related to assessment. At the beginnings, they have to interact with people at the empathy stage, and thus they have to be able to explain what they are trying to do, probably just orally or perhaps with some support of photos or slides in a mobile device. During the project development, students have to show their proposals to the group mates, to the lecturers on charge, and even to final users (during testing stage). Finally, they have to defend their projects in front of the teachers and their own classmates. Thus, all members of the team are supposed to make several presentations, in fact different types of presentations, in front of a variety of persons.

Keep a dynamic attitude and foster an on-going improvement effort.

DT processes put the focus on the proactivity of the participants: it is not a methodology to routinely solve problems, but for dynamically search for the best solution to well-defined problems affecting specific groups of people. Being able to involve the students in the methodology ensures that they achieve this learning outcome. Again, lecturers have to be careful to avoid students to be out of the process.

Consciously carry out the process of the problem defining, including research and analysis of customers and their requirements in the context of the issues being developed taking into account the appropriate tools.

Along the DT stage "definition," each group of students will be able to define the problem, taking into account the needing of the customers known during the research at stage "empathy." The previous workshop provides the students with the tools and methodologies they need to perform this task, and also with some training that will be useful during the development of the project. This learning-by-doing process allows the acquisition of the proposed learning outcome.

Apply the principles and tools of rapid prototyping products and services, and to plan and carry out controlled tests for rapid evaluation of prototypes.

This learning outcome is a consequence of the application of DT methodology, concretely the prototyping stage. The students will learn, and experiment, how to build a rapid prototype of a product, a service or an application, which is useful to allow eventual users to test and evaluate its usability, performance, and usefulness. Thus, a successive round of rapid prototyping and test events allows the students to evolve their designs towards a final proposal. During the workshop, the students would learn the principles of rapid prototyping; and when developing their projects, they would have the possibility of implementing such previously acquired knowledge.

Knowledge of the DT methodology, its application to challenges, and its potential to create highly innovative, market-driven solutions.

Once learning the fundamentals of DT and developing a complete project on DT basis, students would have obtained the knowledge of the methodology as an engine to create innovative solutions, in opposition to standard or canonical ones. This innovation focus is the capital element to have any possibility of reaching the market with the solution: the proposal must be original and innovative.

Additional learning outcomes could be added by local lecturers when designing their subjects based on this course sheet proposal. Here there are a collection of possible additional outcomes, as an example, but each course designer could add those learning outcomes that result to be interesting for the specific objectives:

Plan the development of a team project.

The experience in developing a project working with a group of mates allows the student to reflect on the organization and planning of a team project: the importance of the team cohesion and the empathy with the customers but also among the group members, the definition of the problem to be solved, the different techniques to ideate a solution, the prototyping process and the feedback from eventual costumers prior to improve the solution and present it to the managers. All these steps are, in fact, closely related to the DT stages, being this methodology a good support to that learning outcome.

Become aware of the social, ethical and environmental responsibilities of the engineering profession.

Being involved in human-centered projects, as DT focuses its development, is a good starting point to take conscience of the responsibility of professional activity. Social responsibility springs directly from empathy and testing stages, when students interact with real-world people, "representative" of the society, and take care of the influence or the effect that the proposed project could have on this society. Ethical responsibility is a bit more complex to transmit, or better to experience, but it also bursts forth the contact with people and the guidance of the lecturers and the team mates. Reflections on the ethical implications of the proposals are supported by the collective work performed in teams, and by debates induced by the lecturers when analyzing the outcomes of the different stages along the progress of the project developments. Environmental issues can be easily introducing during the prototyping stage: providing recyclable materials, giving tips for energy consumption reduction, or even forcing the groups to discuss on the greener alternatives to the proposed intermediate solutions.

Implement the knowledge gained through the problem identification.

The capital lesson of the learning experience is to face the problems from the users' view and not from the engineers'. Then, the identification of the problem (i.e., the "definition" stage in DT) becomes the center of the insight acquisition: the project that solves the problem is constructed around it. When finishing the course, the students would be able to take advantage of this problem definition to implement a solution valid for the costumers, not only for the engineers.

11.7   Course syllabus

The course agenda (i.e., the distribution of working hours the students are expected to dedicate) will be organized into three main parts: a workshop, the team construction and the team work. All of them are explained along the following sub-sections.

11.7.1   Design thinking workshop

This section will consist of a short workshop aiming the students to experience the DT process before its application to solve an actual problem in longer-term groups.

The lecturers will involve the students in exercises to experiment empathy, problem definition, ideation, prototyping and testing. The objective is to experience the methodology, not to understand why they do any task or to learn the theoretical basis of DT.

Workshop: space and attitude

The activities during a workshop are different than during a traditional lecture in a classroom. Then, it is important that students realize this fact.

One suggestion is to use a room different than the habitual classroom to conduct these workshop sessions. Ideally, this room has to be flexible (i.e. it does not have fixed desks) and allows the movement of students around tables for team working.

The attitude is also important: many activities must be conducted standing up instead of seat.

The workshop extension will depend on the organization of the local program, but two blocks of six consecutive hours, or four sessions of around 3 h each, with time in the middle, even days, to extend the empathy activity for more than the proposed time, or for doing some team work, would be also an adequate format. Anyway, it should not take more than 5 h.

At the end of the workshop, the students are expected to know the different stages of DT methodology and they would perform a short project on some daily life situation. Examples are "to adapt the coffee shop of the faculty to the users' needing" or "to improve the users' experience of the public toilets at the faculty facilities": the idea is that they begin working in a well-known environment before extending their range in the other blocks of the course. Instead of using large teams, these training projects can be solved by pairs or groups of three people, in order to be agile in the different activities. Dynamics explained along the DT stage chapters can be useful to design the contents of this workshop.

11.7.2   Team building

This second block, the shortest, consisted on the task of creating the different teams, grouping the students of the course to do a coordinated task in the third block. Lecturers will create groups of four to seven students.

There can be many criteria, taking into account that this could determine the future performance of the teams. Among the alternatives, there are: letting the students to group by themselves, at their better preference; or drawing the groups randomly; or by using the results of a simple psychological test done by the students during the workshop, trying to organize balanced teams; or by any other procedure.

At the end of this process, that could be developed in an online session, or even following an asynchronous procedure, each student has to be assigned to a team, and all students have access to the names and contacts of their team mates.

11.7.3 Team work

Each team of students prepares a project providing a solution to a problem detected according to the methodology DT, even in situations of the daily life that perhaps a priori do not relate with their field of study. As DT methodology develops with the following steps: empathy, definition, ideation, prototyping and testing, lecturers will modulate the time assigned to the team work in order to guide the students along the complete process. Lecturers will also take care of the fulfill of all these stages, and provide feedback to the teams in order to maintain a fluid evolution of the project development.

Following DT methodology, the first step (discover and/or empathy step) will be to move the students to put into the shoes of the users (or the employees of a company, or the persons who need a certain service), looking for feeling as themselves. Besides, and depending on the assigned time, the team would search for related news or documents, reports, and so on. Thus, at the end of this stage, a lot of information and feelings would be gathered by the team, being then ready to jump to the next stage.

Then, the team has to define the problem they are going to deal with. At this point, they would take into account that learnt during the empathy step. This second step is crucial, as the definition of the problem guides the rest of the process. Then, lecturers have to be especially attentive so that the team reaches a suitable point of view or, if not, so that they reformulate it until a valid proposal.

During the third stage, ideation, students will pose imaginative solutions and will try to find a proposal that would be reasonable, although it still cannot be implemented given the current technological development. At this point, they will try to identify possible technological or procedural solutions, even using technical and/or scientific information. Lecturers have a mission of questioning everything the team propose, moving the students to think about all conditionings involving in the proposed solution. The mission of lecturers is not to provide a solution or an idea, but to make students uncomfortable with their first solutions, so that they improve on them.

Based on the idea they defined, students will then try to construct a prototype that explains their solution, in the fourth stage. This prototype will be no more than a scheme or a perfectly functional hardware or software object, depending on the complexity of the solution and the time provided to complete this step. If possible, this prototype would take into account legal, environmental, social and sustainability aspects, as thus some elements concerning the learning outcomes are incorporated from the first proposals. The students have to work on the prototype to reflect as much as possible the idea they want to transmit as a solution for the open problem.

Finally, and during the last stage, the prototype has to be tested: it can be done at actual environments, or using the mates as final users. In an ideal situation, the prototype should be evaluated by the targeted user selected in the definition phase. Lecturers can also act as costumers, forcing the situation to check the real world performance of the proposed solution. After this prototyping stage, the team may have the possibility of redo some of the previous stages, in a cycling process, in order to improve the final proposal. Obviously, the time given for that (from some hours to some weeks) will depend on the course structure and timeline.

Each group will document the result of this activity through reports or through an online service such as forum or Wiki. Besides, they will produce a prototype presentation to the end-user. The results will be assessed based on agreed-upon rubrics.

Along all the process, the interaction with the lecturers will be preferably face-to-face, and through forums during the research of information, and by e-mail or online meetings for the exchange of ideas. The lecturer in charge of each team has a mission of being near the team, questioning or giving tips, but not providing solutions or guiding the process.

11.8 Planning

The basic scheme begins with a standard two ECTS course based on DT methodology. As commented previously, depending on local programs, there could be more ECTS assigned to the course, which would be used to go further into the methodology, or going through deeper or more ambitious projects.

Each student will be with the lecturers, in class or seminars, preferable face-to-face for 30 h. Depending on the local program total charge of hours, the students have to dedicate different efforts to complete each part, which will be defined to this time organization: these requirements will define the working hours outside the classroom (or in-classroom without a lecturer), and then the total hours of the course. Obviously, the more hours, the deeper the proposed solution is expected to be (deeper empathy step, more precise definition of the problem, more complex ideation, more complete and functional prototype, and more precise testing and presentation).

So, at the course sheet only the class hours, with lecture interacting with students, will be defined, being the two last columns open for adapting to local programs. Table 11.1 summarizes the organization of the subject, considering four different activities: introductory activities, workshop on DT, projects and presentation or exhibition.

The methodologies used at each activity are broadly explained along next section.

11.9 Methodologies

There are different methodologies to apply during the course development, for students to acquire the proposed competencies. They are explained along the next sub-sections, organized following the activities in Table 11.1.

Table 11.1   Organization of the DT-based subject

Activity	Lect. hours	AW[a] hours	Total
Introductory activities	2	LP[b]	LP
Workshop: DT training	12	LP	LP
Projects: meeting with tutors	15	LP	LP
Presentation/exhibition	1	LP	LP

[a]AW: Autonomous working.
[b]LP indicates that the number or hours depends on local programs.

11.9.1   Introductory activities

During these activities, the lecturers introduce the objectives of the course and some practical hints on DT and team working. This part of the course could be (in fact, it must be) deliver on basis of traditional lesson within a classroom.

The objective of the session is to explain the development of the course, the timeline, the objectives, and what the lecturers expect from students at both individual and team levels. Besides, the team construction would be part of this activity, whatever procedure was selected to apply.

At the end of the activity, the students know what they are going to do along the course and what are the lecturers' expectations on their attitude and performance.

11.9.2   Workshop: design thinking training

Students receive a training on using DT methodology to deal with a given challenge. The lecturers put emphasis in clearly marking the five phases of the process: empathy/discover, definition, ideation, prototyping, and testing.

The teaching methodology has to be oriented to active learning: students have to perform the different activities explained by the lecturers, in order they experience the process of completing all the DT stages to propose innovative solutions to a situation introduced by the professors.

Empathy/discover phase (5 h): The objective is to analyze and understand a given challenge, associating "what to discover" with potential stakeholders, and sketch the stakeholder map (2 h). Once the empathy stage is explained and the students perform different dynamics to improve their abilities in terms of observation, empathic interview and immersion, they are committed to perform a field work. This consists in immersing in inspiring place and interview stakeholders, based on an interviewing scheme articulated around open questions (art of questioning) (3 h).

Define/identify phase (2 h): The students are introduced, and trained in different techniques such as empathy map, affinity map, relationship matrix, or concept map. Then, they have to synthesize their findings (all the information and feelings gathered during the previous stage) using the previously learned visual tools, in order to find insights and formulate the so-called point of view (PoV). Thus, they will have a definition of the problem they try to solve: they learn how to move from a daily situation involving different users to a problem clearly identified.

Ideation phase (2 h): At this stage, different methodologies are introduced to the students: (i) defining "how might we…?" questions based on the PoV to have focus for the ideation session; (ii) brainstorming ideas with different focus points: "Yes, and…" ideas, constrained ideas, analogy-based ideas (d.school framework); (iii) brainstorming-based selection of best ideas: voting, 2×2 comparison map. At the end of the session, students have been trained with these tools, and they also applied those in the project they are developing (i.e. in the solution of the project identified during the previous session).

Prototyping and testing phases (3 h): The methodologies applied in the prototyping include (i) the creation of a low fidelity prototypes using sketching, mock-ups, crafting and (ii) a design workshop where ways to present the prototype are discussed to get it tested by users and get feedback. Applying active learning, students experiment these methodologies at the time they apply them to create the prototypes related to the ideas provided for solving the projects, using the previous phases to support the following. Thus, they construct a prototype for their solutions. Once the prototype is tested by users or customers, they provide feedback on the validity of the solution: some of this feedback will be comments from the users, but some of this must be observed giving attention to the reactions when using the prototypes. Finally, an analysis must be performed: it will consist on a report feedback (studying the agreement between the PoV and the solution), and a reflection on how we can improve the solution and the DT stage at which it would be convenient to go back to iterate again (changes in the problem definition, PoV modification, new ideation process, increasing the quality or the usability of the prototype, and so on).

11.9.3   Projects: meeting with tutors

The methodology of this activity, that is the most time-consuming for the students, is focusing on the team work guided by the tutors. These lecturers will meet the teams periodically, and students' teams will work autonomously between successive meetings.

The proposal is that students develop one larger project in small teams, taking into account the learnt DT steps with few steps of DT. At the end of training, students present their solution to other teams and discuss about their insights.

This is the core of the course: the team of students must address a project, related to an area, situation, or place proposed either by them or by faculty members. During the duration of the course, the team members must work in close cooperation to achieve the objectives of the project. The supervision is such that a weekly one-hour meeting will take place with lecturers in charge. Then, assuming a 15-week duration of the process, a total of 15 face-to-face lecturing hours are scheduled in Table 11.1.

At the end of the course, all members of the team must be able to show its project to the users in oral and/or poster sessions.

11.9.4   Presentations & exhibitions

Every team must present its project to the final users. The oral presentation can be made by one or more members of the team, and must include evidences to show proofs of the work developed and the achieved results. At the end of the presentation, all members of the group must be available for qualification and assessment. The session requires the presence of all members of the team.

When possible, the presentations could be performed during a workshop or a trade show, inviting people from industry and social associations of the University environment to attend it and also to give insight to the proposals.

11.10   Assessment

There are several aspects to be considered for assessment of each student: the selection of the DT tools along the process, the attitude during the different activities and the evaluation of the project itself.

Selection of adequate tools for each specific stage of DT: 20%

The lecturers will consider the tools applied by the team during the different phases of project elaboration, in order to assess the comprehension and application of the previous DT training. This means that the teams must use the variety of tools in order to do forward steps in the development of its project, and lecturers in charge will supervise and mark the selection and the correct application of each method.

Contribution/s delivered during the meetings, exercises, workshops and other activities during the classes: 20%

The flow of the work done during the development of the team project will be marked by taking into account the deliverables the students present to the lecturer in charge, the participation during the periodically scheduled meetings, and the individual contributions to the task.

Project evaluation and presentation: 60%

The project evaluation is proposed to be made during a workshop or conference-like session, in which all students and lecturers attend the presentations of the other projects, done by their mates. During this session, the marks are assigned to each project.

Ideally, this evaluation will be made by lecturers and by peers: 60% (30% done by lecturers, 30% done by peers). Whether there are local regulations that prohibit the assessment by peers, all 60% of the final mark would be decided by the lecturers.

The project evaluation will be performed by means of previously published rubrics, which will be applied by the lecturers and, when possible, by the other

Table 11.2 Assessment rubric for lecturers

Assessed item	0	1	2	3
Group members have been able to gather information related to the proposed problem from different sources. Everyone is a participant in this process.	Scarce search with little initiative.	Part of the group has participated in the search.	Enthusiastic search by part of the group.	All the group has been enthusiastic in the search.
The group has adequately interpreted the problem by identifying its key elements and relating the information collected to them.	No key elements identified.	Key elements not related to the information.	Key elements and some relations with information.	Key elements well related to information.
The group has tackled the problem creatively, providing novel or unconventional ideas and approaches. They have raised little-explored solutions.	No original or novel ideas.	Some conventional ideas.	Some original ideas.	High creativity in proposing original ideas.
After the follow-up meetings and in the presentation of the work itself, it is observed that the proposals evolve based on the comments and reflections on the first versions.	Meetings had no influence on the work.	Some ideas in the meetings, with little influence.	Meetings help the evolution of the work.	Meetings evolved the work and the results.
The presentation of the result shows a maturation of the ideas throughout the work carried out.	Solution is the initial idea.	Solution shows emerging ideas.	Solution is mature but marginally creative.	Mature and creative solution.

0–3 are scores for each item.

Table 11.3 Assessment rubric for fellow students

Assessed item	0	1	2	3
The work shows the ability to face complex problems and to make judgments based on partial or limited initial information.	Team not able to show a complete solution.	Partial and badly justified solution.	Complex problem, but weak justification.	Complex problem and reasonable justification.
The work shows integration of previous engineering knowledge in a broad and multidisciplinary context.	Anyone could provide this solution.	Partial usage of engineering knowledge.	Some relevant engineering knowledge introduced.	A lot of engineering knowledge was applied.
Ability to understand the ethical responsibility and professional deontology of the activity of the Engineering profession.	Ignorance or violation of deontology.	Discussion arises some ethical conflicts.	Solution compatible with deontology.	Deontology from the design of the solution.
Awareness of the need for training and continuous quality improvement, developing values of the dynamics of scientific thought, with a flexible, open and ethical attitude towards diversity (i.e., non-discrimination based on sex, race or religion, respect for fundamental rights, accessibility, etc.).	Evidence of lack of quality principles or sensitivity to diversity.	Neutral solution towards diversity, and some accessibility limitations.	Neutral solution towards diversity, and accessibility.	Diversity and accessibility are included from the design of the solution.
The presentation of the result shows an attractive, creative and reasonable solution to a complex problem.	No clear solution to the problem is shown.	Little creativity in the solution and the presentation.	Solution is not that creative, but presentation is attractive.	Creative and mature solution with a creative presentation.

0–3 are scores for each item.

students (peers' assessment). The rubrics for the evaluation will be published during the first 2 months of teaching the subject, in order every team will know what is going to be taken into account.

Examples of proposed rubrics are presented in Tables 11.2 (for lecturers) and 11.3 (for fellow students).

Chapter 12

Design thinking in higher education: best practices and lessons learnt

Dorota Bociaga[1] and Íñigo Cuiñas[2]

Design thinking (DT) has been employed as a learning strategy with actual Engineering students at different university course levels. During these experiences, lots of situations arise and several solutions can be applied by the lecturers to drive the work sessions well focus on their objectives: to improve the learning experience of the students and to test the applicability of DT basis on improving their problem-solving abilities.

DT fits perfectly on subjects organized as project-based learning (PBL) tasks, and the related experiences of the authors, and their academic teams, are exactly on such kind of activities. Along several years of R/P^{2}BL (research/problem/project based learning) teaching, and after participating in some international innovative education projects focused on DT, a collection of experiences, identified problems, solving strategies, best practices and learned lessons have been compiled, and will be explained in this chapter.

The basis of this chapter is the know how acquired during the collaboration in Erasmus+ international projects, working with students (and lecturers) from various European countries. The international experience in the ambit results to be an additional point, as it allows the authors to identify even cultural or anthropological characteristics that can help when boosting creativity by using DT techniques.

The chapter is organized from different focus (both lecturers and students, but also employers), aiming to provide useful tips extracted from those commented previous experiences.

12.1 How to deliver the DT course?

A course on DT should be focused on transmitting the students how they must do for implementing a process of developing a project that solves, or at least mitigates, a problem, in fact a real diagnosed problem. A clear difference with other project

[1] Łódź University of Technology [DT4u], Poland
[2] atlanTTic, Universidade de Vigo [GID DESIRE], Spain

methodologies is that DT puts the human in the center of the process: it is not the main goal to solve an engineering problem, or a scientific problem, or an economic problem – those are incidental to meeting the needs of people/end-users. This specific characteristic is also a factor determining the best approach to deliver the course to the students: whether the introduced methodology is human-centered, the students should have contact with humans, i.e. people with implications in the learning projects, to understand the special connotations and to practice the application in actual situations.

This previous thought links directly with the delivering methodology. Teaching a DT is different than other type of methodologies, which can be explained based on their subjacent theory. All around a project developed on DT basis focus on the people. So, a complete training in DT needs the connection with people and experience the unexpected discoveries that a design-thinker can extract due precisely to this contact with humans. Then, learning by doing arises as the best strategy to deliver a course on DT.

The students have to know the theoretical basis of DT, of course. But these could be introduced in parallel of the experience of performing a challenge involving real-world people: both proposing different activities to develop the project and explaining why the lecturers are asking students to do each activity.

The capital element is to acquire experience in contacting people. Students have to test the unexpected world of empathy with persons that have different feelings, different backgrounds, different lives...in fact, they have to learn that there are many ways to observe the world that perhaps they never thought. Discovering the richness of others' lives, and new points of view, must to be lived, not to be taught.

What is the best way to deliver the DT course? – step by step, letting students to discover the methodology stage by stage but mostly, letting them to discover how they can collaborate as a team, how much empathetic they can be, and how much they can immerse in the situation to use their best skills in order to build solution most accurate adjusted to the people's needs.

12.2   The DT stages and some tips for each of them

When training students in DT, each of the stages presents specific situations or difficulties that must be considered. In general, empathy used to be the most time-consuming block, definition is crucial, ideation must be carefully supervised, prototyping have a lot of fun for students, and testing is many times forgotten or only partially introduced. Anyway, there are some tips that could help when training students in DT methodology.

During the empathy stage, the temptation of substituting the direct contact with users by an online questionnaire is always present as a way of reducing the time of this action. Perhaps one could think that the questionnaire is more efficient, and it is effective when the objective is to gather a number of data for clearly standardized issues. However, a closed questionnaire will never provide more than cold figures: it is really difficult, or even impossible, to extract sensations, or feelings. This means that it is really difficult to empathize with users if there is no direct contact with them.

Students have to experiment interacting with people; interviewing them; extracting their interests, feelings and motivations; observing their attitudes; and living within their shoes in order to really perform the empathy stage. In fact, empathy is possibly the stage that needs more time to be experienced, as it deserves some guided training in observation, concentration, and interviewing, and some more free-running time for immersive experiences within real-world users. A strong empathy experience will usually lead to solid foundations to construct the rest of the project on them. Thus, this stage deserves many care and time with students.

Online questionnaires could be indeed useful to confirm or to extend the insights extracted from a number of interviewees to a broader universe, but not to substitute the empathy stage. They could be identified as additional information. Using a shortcut at this initial moment could compromise the development of the full project.

During the definition stage, a typical temptation is to advance in the solution to the problem in parallel to define it. Lecturers have to be clear and to take care of the process: when students are defining the problem at the same time they are trying to solve it, probably they focus on defining a problem that fits with the solution they have ideated in advance, and not with the insights extracted from the empathy stage. Students also want to find a solution, as they feel more comfortable when they are seeing the light at the end of the tunnel. Nevertheless, lecturers should control and limit this temptation: we are not looking for "our" solution, but for the problem discovered after the immersion with users. Thus, lecturers must keep the focus even when students want to run faster.

Ideation also has some difficulties, regarding the conservative character of students in general, and engineering students in particular. Even they are commonly younger than lecturers, many times it is incredible how traditional they could be when looking for solutions to a problem. Assuming that the definition of the problem fits the users' expectations, lecturers must be aware on the proposals arising from each students' team. The clear temptation at this stage is to look for a canonical solution, or a collection of canonical solutions, which application is well known by students or they could find it in books or the Internet. Education programs have been designed to train persons that apply very efficiently a catalogue of solutions for standard problems. Thus, our students are well trained to find these standard problems, or to divide a complex problem into several standard problems, for then proposing a solution or a concatenation of solutions previously learned. Creativity has not been in the center of our programs, and many times our students have limited creativity to some subjects (i.e. arts and crafts) but they do not consider its role in more technical, scientific or even social ambits. Supervising the process and forcing students to explore nontypical or non-canonical solutions is the main task for lecturers during ideation stage: in other case, the quality and innovation of the final proposal could be compromised.

Prototyping is very popular among students: they have fun constructing their proposals and they realize that they are able to create something tangible, something that can be touched and that represent their own solution to the defined problem. For them, it is the top of the mountain, the end of the project. Sadly, sometimes it is actually the end of the project! At this stage, students (and principally, lecturers!) have to remember that the aim of the methodology is to satisfy the users needing, not to provide "our" solution to the users' problem we have defined. Prototyping is just one step, another step, in the way to improving users' experience. It is natural that students forget this, as they are engaged in the manual work of providing corporal shape to their ideas. But lecturers should keep their feet on the floor, and constantly remember that the prototype is not good or bad, valid or useless, by itself. The goodness of a prototype depends exclusively on its performance during the interaction with users. Lecturers should not assess the prototypes that students created, as prototyping is not the final objective of the project design.

The actual end of the project is testing stage. At this point, users interact with the proposed solutions whereas students (designers) observe how they manage the prototypes and also interview them to know if the answer really solves the detected problem. Testing is interesting to learn about possible improvements and adjustments before creating a final product to be sent to the market. Thus, testing is as important, or even more, than previous stages...although many times it is avoided. The temptation in this last stage is to manage it as the concluding remarks of the course, limiting it to a succession of presentations following a conference session scheme. Of course, moving people outside the course that could act as users to test the prototypes is more complex than using the class mates as mock-users. However, without a deep testing stage, the DT process is not complete, as the human-center condition is lost when nobody is testing the prototypes. When feeling this temptation, lecturers should remember that the key element of DT is that people are at the center of the process, and thus they have to select a group of people that can test and provide feedback on the prototypes.

Is testing really the end of the project? Did the students (and lecturers) learn anything about their solutions during the interaction with users? Is there anything in the solution that should be improved with the new insights gathered during testing stage? Or, perhaps, do students think that they should re-define the problem, and/or re-ideate the solution, and/or re-prototype their proposal once users have interacted with the first prototype? Many times, the end of the five stages is no more than an open door to a new round of activities to improve the proposal, or even to begin a new solution. And this is because DT is not exactly a linear process, or a concatenation of recipes, but a methodology to enter into a loop of continuous approach to the best solution. Of course, this is not always possible in an academic course, which flexibility in dedication and duration would be limited by several norms and traditions, but when possible lecturers must boost in their students' minds.

12.3   Tracking the teams

Due to the eminently practical orientation of the methodology, DT training should be designed on the basis of "learning by doing." Even more, students must be grouped in teams, as DT involves different activities thought to be developed collectively, in cooperative work. Students should feel that team working is the only way to do DT, i.e., they have to realize that tasks performed on DT basis could not be developed individually. This team-oriented focus is one of the tasks of the lecturers, and they have to cope with this idea providing constant supervision. Besides, this supervision will help in assessing the learning process: at the end of the training period, we have to be sure the students have learned how to apply DT basis to solve a human-centered problem.

During the DT process, the teams are expected to evolve in their activities, reaching different milestones at least at the end of each of the DT stages. Lecturers in charge have an important mission of supervision, but also of assessment, as the tasks are going to be incorporated to a subject in the academic program. As DT process is a succession of activities towards a solution to the identified problem, assessment should be performed in a continuous evaluation methodology: we are interested in what, but also in how. The final proposal is a "what," it is a solution to improve the users' activity or experience in some ambit. Nevertheless, the "how," the progression on the project development following the different DT steps is as important (or even more) as the final solution, as the training objective is not to find a solution but to learn how to provide solutions using DT methodology. As a consequence, lecturers have to follow the process and to assess each step, in order to move the students to advance in the route without jumps to avoid any of the steps.

Empathy can be supervised with a team meeting during which the students could explain their activities to immerse into users' world: photos, interview notes, and also oral descriptions of what they did. This needs an additional effort of the lecturer to receive the information, to understand what students did, and not to guide the next steps.

Definition is probably the easiest step for supervision, as its end is a clear statement commonly identify as the point of view (PoV). Thus, a short report of the team with the PoV sentence would be enough to follow the activity.

Supervising ideation requires the presence of the lecturer, as an observer, during the main brainstorming session when the members of the team provide ideas for a solution to the problem defined by the PoV. After this activity, one member of the team should write the minutes summarizing the process and describing the selected ideas. Prototyping could involve several hours of team work, most of them without direct supervision from lecturers. However, it is recommendable to prepare a short presentation to show the advances to the team, and to show the evolution to the lecturer in charge, in order to receive external feedback and also to generate assessment evidences for marking the subject taught under DT basis.

A testing event must be organized to check the solutions. Ideally, users involved in empathy stage could participate in this activity. At least, students from the rest of the teams should interact with the prototypes to check how a non-participant person faces the proposals and to decide the improvements and changes there should be added to the first attempt of solution. Each team must write, at the end of the event, a short report on their prototype performance, and the lessons learned after the testing event.

After this testing event, a meeting of each team with the lecturers could be interesting to decide the changes to introduce to the prototype to reach a final solution, which should be presented to the lecturers' panel after some additional time to improve the team work.

In summary, supervision is needed along the full process of DT-based team working, but each of the stages would require different strategies in order to provide efficient support, and to gather good evidences for assessing the academic performance.

12.4 Course delivering

The planning of the DT course must be defined as a function of the available time with the students and the concentration of these hours in days, weeks or months. Many possibilities are open, and it is possible to provide the students with DT skills in all of them, at different depth depending on the format, and also with different implications and autonomous work. Along this section, we are going to analyze how we could deliver DT courses on one day, one week or one semester, analyzing its advantages and disadvantages.

12.4.1 One-day course

When the objective is to introduce DT methodology in one-day course, we have to focus in the acquisition of the fundamentals of DT in a practical way. We should not expend a lot of time in theoretical dissertations, as we do not have time! The proposal is to provide hands-on sessions, intensive and guided to move the students to experiment all five stages of DT, perhaps introducing only basic tools (or a limited number of them) and challenging the students with some proposals that they could identify as really close to them: i.e. the challenge could be "how to improve the users' experience in the public toilets at your faculty or school," or "how to improve the users' experience in the cafeteria." The idea is to allow both empathy and testing stages with real-world users, but keeping the distances in a controlled range not to extend too much the time required for the different activities.

The proposal is a combination of activities within a classroom, in workshop format, and free-running guided tasks to perform the empathy stage by teams of four or five students. The time within the classroom should be mainly focused on moving the students to experiment different tools to train them into the different stages: the focus will be put on the "how" more than on the "why."

One-day course

The proposal is a workshop with up to 6 h of face-to-face activities and 2–3 h of autonomous work in teams:

1. Workshop #1 (2 h)
    (a) Observation.
    (b) Interaction.
    (c) Immersion.
    (d) Challenge introduction.
2. Team-working (2 h)
    (a) Autonomous empathy time.

3. Workshop #2 (4 h)
    (a) Definition/synthesis
    (b) Ideation
    (c) Prototyping
    (d) Testing
    (e) Closing remarks

Thus, we suggest to use the initial 2 h (or one hour and a half) in guiding the students across different activities to improve their empathy skills: introducing themselves by means of a game, promoting the observation by drawing, and interviewing or just looking in pairs. Then, the lecturers provide some general explanation on the DT methodology, its five stages, and the challenge to be developed. This challenge should be limited in space, being somewhere not far from the classroom or seminar, and also in time, related to the students' usual activity or environment. This section can finish with some tips for doing an empathic interview, providing some time to the teams to create their strategies for the interview time.

The next 2 h are devoted to autonomous work performed by the different students' teams in order to observe and interview different users of the environment selected for the challenge. During this time, the lecturers should observe how the students interact with real-world users, in order to provide feedback during the next face-to-face session. However, there is no time for lecturers to participate.

After this empathy time, the largest part of the workshop takes place. Now it is time to involve the students' teams into different learning-by-doing tasks, introducing them successive tools to develop the DT stages: definition, ideation, prototyping and testing. Among these team activities, it is important to do some assembly events to share the evolution and the feelings of the students. At least, a team presentation of the points of view, the selected ideas and, then, the prototypes are suggested to be included along this session.

12.4.2   One-week course

One week allows the lecturers to provide a deeper training in DT methodology, combining the practical learning (the "how") with the theoretical or methodological basis (the "why"). Dividing the course in five sessions, each of them can be devoted to one of the DT stages: empathy, definition, ideation, prototyping and testing, leaving additional "homework" for the teams to perform empathic interviews out of the workshop timetable, or alternatively it is possible to use the first two of the days for empathy and combine definition and ideation in the third session.

Empathy deserves at least a full session and some additional time of autonomous work by teams. During this time, the students would be introduced in the DT methodology, trained in empathic immersion and committed to perform a challenge that could be more ambitious than that suggested to the one-day course: the environment could be larger and further away (i.e. a mall, a railway station, etc.) and the empathic research is expected to be deeper.

After the autonomous time, the session devoted to definition has to be a combination of theoretical items (not only the tools but also why we use these tools, which tool resulted to be more adequate for each situation) and to hands-on work: the teams of students must use the information gathered during the empathy activity to extract the insight from them and to conclude with the definition of a PoV, as a base to construct the rest of the process.

The next day, the task would be the creation of ideas that respond to the defined PoV. Students will need some explanation on different ideation techniques, and also training on those techniques. Then, they will use some of them, by teams, to provide the idea for solving the identified problem.

Having the basic idea, the fourth session is devoted to creating prototypes representing this idea, and to evolve the original idea taking into account the interaction of the design team with the prototype. At the beginning of the session, different prototyping techniques must be introduced and students' teams have to work in the development of a first version of their prototypes.

The last session begins with the introduction on how to organize a testing event with users. A mock session is then performed, with each team presenting their prototypes and adopting the rest of the students the role of users. Then, the teams will use the feedback to improve their prototypes for some time and, finally, a session with users, with real-world users, should be organized as a closing activity.

To sum up,

One-week course

The proposal is a workshop with five sessions of face-to-face activities and, if possible, another session of autonomous work in teams:

1. Day 1. Workshop (4 h)
   (a) Introduction and presentation.
   (b) DT fundamentals.
   (c) Design
   (d) Empathy
   (e) Challenge introduction.
2. Day 1. Autonomous work (4 h)
   (a) Observation
   (b) Empathic interviews
3. Day 2. Workshop (4 h)
   (a) Definition
   (b) Mind maps
   (c) .
4. Day 3. Workshop (4 h)
   (a) Ideation
   (b) Ideation techniques
   (c) Active brainstorming
   (d) Storytelling

5. Day 4. Workshop (4 h)
   (a) Prototyping techniques
   (b) Low-definition prototyping
6. Day 5. Workshop (4 h)
   (a) Testing events
   (b) Prototyping improvement
   (c) Testing with real-world users
   (d) Closing remarks

12.4.3 One-semester course

Delivering a DT course along one semester has the advantage of providing the possibility of performing a project that is closer to the real world activity. Students' teams have time enough to do successive refinements in their proposals, being involved into an iterative process in which all DT stages could be reformulated to improve the final solution proposed to a panel of final users. This format fits perfectly on a PBL style topic.

The proposal should be divided into two parts: the first one focus on learning the methodology and the second on developing a project, providing solutions to improve the experience of users of certain environments or services.

The learning part could follow the schemes of the one-day or one-week course, depending on the time scheduled in the academic organization. The idea is to provide DT tools to the students' teams in a concentrated term, ending this training period with the introduction of the challenge. The lecture staff must reflect in depth on the challenge proposed: it must be complex enough, including social empathic interaction but also research on the topic, and they also must define what they expect from the prototypes (just the scheme and description, a semi-functional or mock one, or a fully functional solution).

One-week course

The proposal is a combination of a learning part in workshop format and an autonomous work part, with some control points along it:

1. Workshop. DT training
   (a) Scheme of one-day or one-week proposals
   (b) Challenge introduction
   (c) Calendar agreement
2. Team-working. Solution of a project
   (a) Empathy stage
   (b) Definition stage: point of view
   (c) Team meeting with the lecturer in charge
   (d) Ideation stage
   (e) Team meeting with the lecturer in charge

    (f) Prototyping
    (g) Testing event with possible users or, at least, with the rest of the students' teams and the lecturers
    (h) Iterative process regarding the feedback, jumping back to any of the stages from (a) to (f)
3. Assessment event

Then, students have some weeks to perform the different DT stages in an autonomous work, supervising by the lecturers in some milestones along the provided time. This calendar should be discussed and agreed by both lecturers and students.

At the end of the process, an assessment event has to be scheduled to provide academic feedback to the different teams and to conclude the semester.

12.4.4   Teams: main issues

DT places its foundations on having a group of people working together in interdisciplinary teams when possible. Building teams and managing them in a proper way are the basics to the success of the process. The size of the team, how to build it and to share the tasks, and how to promote the individual potential and relationships to improve the team performance are open issues to cope with along this section.

When dealing with a group of students joining a PBL topic based on DT, a question on the size of the team will probably arise. And it has not a direct solution. Small teams (up to five members) are faster in solving problems, but they are not efficient when a very wide empathy strategy is needed: when a lot of interviews are required, the small teams require a lot of time to complete the work, and this limits the efficiency of the results. On the other hand, very large teams (i.e. from 11 to 19 members) are well-suited for gathering information but they suffer important problems of communication, and group meetings become long and unproductive: then, managing the gathered information to define the problem and ideate solutions could represent big problems. So, a balance is required and a number up to seven members seems to be ideal for combining efficient gathering data and fast processing of those data to define and ideate. Groups of this size keep the cohesion among members, which mutually stimulate during the different stages of the process.

Besides the size, other issues should be taken into account when dealing with team building: the members' diversity, their trainings in teamwork and the cohesion among them are capital for the performance. The diversity in terms of knowledge level, experience and seniority has been demonstrated to have positive effect on the evolution of the team. In general, homogeneity is not a recommended criterion for building the team. Independently of the previous knowledge of the teams, it is important not to assume that the members have training in teamwork skills; you must be sure that they have and, in other case, that you can provide some basic training to make them easy the interaction and collaboration. This common training is a way to increase the team cohesion, as well as organizing group activities that force the interaction. Cohesion helps in supporting the well development of the common task, even though it could be limiting when the cohesion is too high and the productivity could decrease because the group has other interests.

Regarding the tasks assigned to the team, they must be adjusted to its potential: when tasks are low level compared to the team potential, the members lose their motivation; but when they are expected to solve too difficult problems they could be frustrated and do not reach the requirements. A good tip could be to set goals with increased difficulties involving the entire team, feasible goals, motivating the team members to increase their performance in order to successively reach each of the challenges. Besides, it is useful to highlight the positive impact of the matters discussed and agreed during the meetings, having the lecturers a chance to guide the process when the task is not advancing. Remember that guiding is not the same as providing solutions.

Analyzing the individual potential of the members is also interesting when building a team. This means that both personalities and having complementary skills have to be considered. The performance of the work increases when adding extraverted and open personalities, which improves the comfort among the team members and gives chance to all of them to be deeper involved. Having complementary skills produces positive impact on the effectiveness of the work as these skills add instead of deduct in the operation of the team.

An additional action of interest is considering the individual relationships among the members of the team. Establishing strong and healthful relations promotes the joint responsibility on the team work, moving the feelings from the individual and partial responsibility to a global one.

12.4.5 Teacher's main role

The role of the lecturer when delivering DT workshops or training sessions has two main issues to take into account. One is related to the strong focus on "learning by doing." DT is not a methodology that needs students to understand and learn a lot of theoretical concepts, but it needs to be experienced. Students are not to be taught but to be guided along a process, both personal and from third persons, realizing that what it is being proposed is true, real, and works in practice. The other issue related to the implication is of more than one lecturer during the training sessions. At least two persons must act as teachers, one acting more as an instructor that introduces the topics and provides instructions and the other one as a supporting tutor that gives assistance and helps to reinforce the main ideas. He also provides additional experiences and has a clearer view of the classroom atmosphere, complementing and providing different views and experiences about the contents. From these two premises, teachers could adopt different roles during the DT teaching and training. It is worth to note that these different roles can be experienced in a progressive way while the training evolves from the first sessions towards the end of the process, and selecting the right role at each moment, depending on the students' team evolution, would be helpful for the whole experience. Those roles could be:

Lecturer: A lecturer offers the topics to be learned and proposes activities to the classroom. This role is closer to traditional teaching, although even in this role, the teacher is mainly acting as an experienced person that has a certain knowledge and is convinced about the benefits of the method.

Leader: A leader inspires and motivates the students, and the teacher must try to do that when in this role. He also introduces the different stages and techniques involved in the DT methodology, and specially shows and reinforces the key ideas, such as empathy, testing or feedback.

Tutor: A tutor provides assistance to students on the performance of tasks and tries to help those that are struggling in. To a long extend, the tutor exists to serve the wishes of the students on the performance of their tasks. Nevertheless, in no way, the tutor is the principal actor, but a secondary one. The tutor lets the students to work by themselves, to carry out the proposed tasks, to decide about the resulting outcomes, etc. As a tutor, the teacher just acts when the students are blocked or when they ask for support.

Mentor: A mentor is a person with experience in a certain topic that guides a less experienced person by offering indications from the personal experience and expertise. In DT training, the teacher does neither offer the good answers nor provide direct instructions, but listens and inquiries the learners in order to get the actions and answers out from themselves. From a teaching PoV, this mentor role is aligned with the goal of empowering the learners as DT researchers that are able to proceed with the process on their own foot.

Supervisor: A supervisor just superintends what is happening during the DT training sessions. In this role, teachers are almost invisible, they just go around the class-room making observations about what the students are doing, but without taking part or providing any feedback. Most of the time, if students are working in a right way, teachers will be acting of this way, collecting data, sentences and observations (e.g. pictures). All these elements are very useful to be shared later with the students. Usually, at the end of the day, all the teachers and students make a circle around and talk about the whole day, sharing and reflecting the main lessons and experiences. The annotations of the teacher as an observer are very powerful at this point to reinforce and empower the students themselves as real design thinkers.

12.5   Managing interdisciplinary teams

Interdisciplinary teams are advantageous compared to those very homogeneous in terms of members' education and knowledge. Having different origins and different academic and vital knowledge, the ideas and proposals will arise from different perspectives, allowing an eye-opening process in all the members. When facing a problem from different angles, difficulties seem to be less, or at least they are possible to be attacked with a variety of proposals. From this optics, the interdisciplinary character is a must be when building a team, and when possible we should try to take that into account.

Besides, the variety of knowledge could provide better comprehension and a wider range of ideas to solve the identified problem, as technical solutions would arise from different specialties.

However, there could appear some organization problems due to having such different profiles within the same group. One is the different cultures of work among the members, which are probably used to work in a typical way of their own areas. Solving that would require efforts and comprehension from the side of all members, even the empathic characteristics of a design-thinker should help in overpassing that situation.

Other problems are related to the logistics. Universities often organize their courses by areas of knowledge, and many times even the buildings could be really far away. For example, it is not usual to have the Humanistic Studies in the same building of Engineering; or Nursery and Economics sharing the space. In big companies the effect is similar, as they group persons by departments, and the main trend is to group departments in buildings. The logistics to meet the members of a team, i.e. the time they need to be in the same place at the same time, as well as the difficulties in fix the calendars of a variety of faculties or departments, represent a not minor limitation in the team management.

12.5.1   Students' perception on DT

Students generally perceive DT as a novel approach, as something different to mainstream techniques to face problem solving. It helps to promote creative confidence, to get away from the *strict* and *narrow-minded* ways of facing problems and their solutions. After completing the educational experience, students understand that the design thinking methodology can be applied to complex situations beyond the design of new artifacts or services, involving a community of interacting users with different visions and requirements.

As a side contribution, the courses demonstrate that the DT methodology is most appropriate to address problems requiring a user-centered approach by ad-hoc multidisciplinary teams, as it dramatically facilitates group building and coordination among solution designers with different backgrounds and expectations. Incidentally, students explicitly value the participation in multidisciplinary teams. Indeed, design thinking can play the role of a cross-disciplinary methodology, which allows a team across disciplines to develop a shared understanding of problems and solutions, as it broadens disciplinary reasoning and helps to forget about established internalized along their university studies.

Skills developed by design thinking include working with people and in teams; being creative and innovative in the workplace, confidence, presentation skills, analytical thinking, and dealing with difficult people by putting into practice soft skills such as empathy, active listening, and positive orientation. Students become especially aware of user orientation and, related to this, the importance of empathy to understand the needs of real users. However, in some cases, it is difficult for students to understand that interpreting the needs of users is not just about deducing, but also included a relevant amount of sense making.

Students also have some difficulties with their full integration in multidisciplinary teams. One of the DT aspects that is more difficult to transmit to the students is that "the team is greater than the sums of individuals."

Experience also demonstrated that it is not easy to overcome a strong attachment to the solution chosen (i.e., it is hard to go back and discard some solution that seems to work but is far from the being the most convenient solution from a DT perspective): students tend to defend their proposal as a personal question, instead of assuming that any solution could be improved, or even that any path to reach a solution could be walked back to a confluence and then select a different direction.

Although courses offered by the authors are in general perceived as useful and positive, in a limited number of cases, the course did not met the students' expectations, being in most cases engineering students the ones having this perception. The main reason for this seems to be that problem-solving approaches in design (i.e., the seminal field of application of design thinking) are different from that in engineering. Whereas DT allows dealing with the ambiguity of design problems as wicked problems, the mental processes of engineering students is typically more biased to the development of effective technical solutions. In the case of problem-based-learning projects in engineering, the difficulties in communication between experts (i.e., engineering students) and non-experts (i.e., final users) during the engineering development process make the blend of both approaches complicated and lead to a dominance of analytic–systematic approaches to problem and solution finding.

It is also possible to observe a trend to perceive DT as something difficult to apply within existing organizational structures in companies and organizations, no matter that it is generally appealing to developers when it is communicated during courses and other educational initiatives. Related to this, a kind of risk perception can also be identified. For example, developers, designers and practitioners accustomed to standard procedures in engineering, economics or science see as a real challenge to match existing performance or outcome indicators and project milestones with the empathy-based and explorative paths of DT. In other words, DT is in some cases perceived as a fuzzy, unclear approach to project development that may help to think out of the box, but also may compromise communication of results and justification to customers and other stakeholders.

One approach to face the situation above could be to identify specific tools to assist the developer in implementing DT, to make them to identify DT with an adaptive toolbox including tools that can be applied depending on what kind of problem they face along the development process. This would make DT more flexible, becoming something that can be applied on demand and in a gradual way.

12.5.2   *Employers' expectations on students educated in DT*

Quality employment depends on many factors, being some of them the skills acquired by prospective graduate (i.e., to-be-employees) that would provide a competitive advantage during job-seeking. According to many companies, it is an outdated approach to focus on professions that may be popular in the future. Rather we should concentrate on competencies. The analysis of the material gathered during interviews with various employers in Europe as well as other commercially available data indicates that for most professional categories, employees put special emphasis on self-organization skills. The exception is people working in the service industry who should have good interpersonal skills to ensure good customer experience. Specialists and engineers should have good social competence. These skills have particular significance as they are hard to learn or acquire and it takes a long time to do so.

Employers are aware that employees will need to undergo training of the so-called "hard skills" required by the specification of the workplace or position. Research clearly shows that professional competencies are important in the case of managers, laborers, operators, or assemblers and less important for office workers or specialists. An in-depth analysis of employers' recommendations and job offers made it possible to identify detailed interpersonal and self-organization competencies important for particular professional categories. Managers, regardless of professional category, should have good communication skills, show their initiative, be independent and good at time management, and be able to work in a team. As for people working in the service industry, additional emphasis is put on ease in establishing and maintaining good contact and relationship with the customer. Specialists should have good communication skills, be independent and have good time-management skills. Teamwork and pro-activeness were also stressed here. Being pro-active is also required from office workers and technicians besides independence and good time management skills. People employed in the service industry should have good communication skills, be pro-active and sellers ought to additionally have ease in establishing contact with the customer. In the case of laborers, qualified or unqualified, assemblers or machine operators' interpersonal competencies are much less important. They should have self-organization skills such as pro-activeness and a taste for entrepreneurship. In these professions, professional competencies are the most important and often certificates shall confirm them.

To sum up, on the basis of the data gathered, 10 such competencies were selected:

- The ability to extract hidden meanings, interpretation of hidden content, understanding of facts so important in the process of decision making.
- Ability to communicate with others on emotional level, social intelligence.
- Ability to think and come up with solutions which are not governed by strictly defined rules: reaction to non-standard situations, using creativity.
- Ability to work in various environments, the so-called "multicultural competence." The best team is a team whose members vary in age, skills, way of thinking and working. It generates many outcomes and facilitates detailed assessment of possible solutions which helps in choosing the best one.
- Ability to process big portions of data into abstract concepts and understand proofs depending on this information, the so-called "analytical thinking." Sought after are people capable of performing statistical and quantitative analyses as the body of available data is ever growing.
- Ability to understand and read various forms of information, video or picture and being able to analyze them and interpret. Needed are people who are skilled in reading and creating these types of messages, capable of communicating one's work results not only in the form of documents or presentations but also video.
- Understanding concepts which span multiple disciplines, that is, interdisciplinary competence. Sought after are employees who have a specialized knowledge in one

particular field complemented by general knowledge from other fields necessary
to solve a complicated problem.

- Ability to develop work's tasks and processes in such a way as to ensure reaching
  its goals, the so-called project-oriented approach. It also means ability to change
  the environment of work so that it has a positive effect on project's execution and
  finalization.
- Ability to filter out information. Selecting only information important for task
  completion.
- Ability to operate with due commitment and be part of teams regardless of the
  present location of their members, on-line cooperation.

It can be noticed that the above competencies are a response to the changes
in the contemporary world. Companies these days operate largely on international
markets, thus their employees should feel comfortable in international teams. For
that the ability to speak foreign languages and use mobile technologies is needed.
An employee should be multi-functional, capable of accepting frequent changes to
the scope of their responsibilities or even position of work. It is important to be
flexible or display willingness to retrain as assigning new tasks to employees is not a
rare occurrence in small business particularly but it is true of big companies as well.
Knowledge and IT technologies management skills are no less significant.

12.6   Lessons learnt

After several years using DT in university courses at different levels (from bachelors
to masters, but even in research projects), the experience allows us to identify what
is commonly working well and what does not run anymore.

In general, students are very proactive in empathy exercises at classroom, involving the other mates, as they have fun doing such training activities. However, they are
lazier when empathy stage is asked to be performed autonomously and, mainly, when
this activity extends for several days (or even weeks): in this case, the initial enthusiasm
decays progressively. Thus, strategies to maintain the intensity along all the empathy tasks should be included: i.e. intermediate milestones, conquest regarding the
amount of interviewed people (or, better, the amount of gathered information during
the immersive actions), attractive environments to perform the empathy immersion,
or any other that could engage the students to the task.

Students, and also professionals, could have the temptation of using surveys
instead empathy immersion. This is natural, as it is less time consuming and more
comfortable than a deep empathy stage, but this practice perverts the DT process
itself, as most of the human-center focus would be fuzzy with this strategy. Of course,
surveys could be interesting supporting data, as reports, scientific papers, or other
sources of information, but they must not replace the direct contact with users. So,
a best practice could be the direct supervision of the empathy process, to force the
students to effectively apply the full methodology.

During the definition stage, a good practice is to keep all time some lecturers supervising the team working process and, if the case, reformulating or providing some tips to drive the evolution of the session toward a good PoV. We learned that allowing a too autonomous work during this stage, in a training process, could blur the rest of the development.

We should encourage students to be very crazy during ideation, or at least not be too canonical. Students enjoy in a relaxed environment and new and original ideas would arise when providing the ideal time and place. Guiding by lecturers should be reduced or eliminated during this part of the process, but supervising is required.

In between ideation and prototyping, some help from lecturers should be needed, at least to check the ideas and to be sure that the team extracted all the insights from the gathered data and the PoV. The minutes of the ideation session are useful for this milestone.

Student teams would need some time for prototyping, and autonomy is a must in this case. However, some intermediate meetings (or short contacts) with lecturers would be interesting to know the evolution, to help in solving problems, and also to check that all the team is involved in the process (even for assessment).

The experience tells that many times testing is the Ugly Duckling among the DT stages: it is the last one, sometimes is done in few minutes, and difficulty with real-world users. We learned that doing testing events helps a lot in getting insights on the prototypes' performance, allows the teams to improve their solutions, and gives good information to lecturers in charge in order to provide valid assessment to the work done.

As a general tip, we learned that students are better trained when they experiment all the details of the methodology, and they learn more when they have fun during the activities.

Chapter 13
Design thinking for boosting creativity
Jae Ho Park[1]

Design thinking (DT) is philosophy, mindset, and strategy which helps people to reach the creativity and innovation. DT is like a compass and a methodology to guide sailors to sail a ship in the uncharted ocean of creativity. However, in order to boost creativity for entrepreneurship we need steps as follows: Curiosity, Imagination, and Innovation.

We human beings have confronted the pandemic challenges in every 100 years since the history of human beings. These challenges were Pest, Cholera and Spanish influenza and COVID-19. To solve these problems and confrontations, we are in need different ways of thinking which is based on Creativity. The challenge is a Chance. Chance has two meanings in Asian cultures: danger as well as opportunity. If we focus on positive side of chance, we could boost creativity so as to find out new solutions.

DT is a methodology to find out the need of customers (needs finding), the definition of an actual problem, idea generation through communication and collaboration. It also permits to make many mistakes which is inevitable in the process of learning. The last stage of DT, customer involvement in the form of test, suggests new ways to find a solution.

This chapter is devoted to introduce DT cases with healthcare organization OD (Organisational Development), train the DT trainer processes, career development planning for graduate students, and cases of leadership development in industrial organizations.

13.1   Learning Creativity and innovation from American History

Let's learn from Edison for the sake of national reconstruction! It was a slogan for social innovation movement in the United States in 1897s. Thomas Edison, who was called *Napoleon of the Invention* and/or *The Magician of Menlo Park*, was a superman who raised nearly 10% of the National GDP, the wealth of American society, through his more than 1,000 inventions. Edison was *A guide to the future* and led his era as an *idea pilot* by his deep reflection, creativity and innovation.

Thomas Edison invented the light bulb and at the same time he also developed the entire ecosystem of electricity-related industries. In other words, Edison's light

---

[1] Department of Psychology, Yeungnam University, South Korea

bulb was known to the public as a representative of his inventions, but if he would not have developed the power generating system to generate electricity and a transmission system to send electricity, the light bulb would have been just an invention to replace candles.

Edison's genius lies in the fact that he did not simply invent a single product. In addition to that, he addressed simultaneously the entire market-associated total system with that invention. Edison has always predicted where and how people would use the product of his inventions. And this prediction precisely matched customer's needs. Furthermore, he also modified an already existing inventions and transformed them to different products in accordance with the need of users.

Of course, his foresight has not been always correct. For example, he expected the gramophone which he has invented, would be mainly used for recording and playback of communication as a recorder player. However, the customers used it as an innovative device of music listening. He always thought about what is the need of customer/user and the preference of market. He also contemplated what would be the future need and motivation of the users.

In order to find out customer need, he read enormous amount of materials every day. He read several kinds of newspapers, weekly- and monthly magazines. His well-known speedy reading technique helped him so that he could digest such an amount of reading materials which he used to capture the new trend of society as well as the expectation of people. He approached his user and client with the attitude of customer satisfaction/customer touching which became popular nearly 10 decades later. This became popular nearly 100 years later.

Edison regularly wrote articles and columns in the newspapers and magazines about what he would expect to come next. This means that he used mass media to lead and guide the need and motivation of technology transformation for people.

Edison-style R&D corresponds to the origin of the creativity-methodology which is nowadays called *Design Thinking* (human-centered innovation method). If we look at Edison's research cases, there are always certain procedures that must be preceded before the research management method in which a specific product is made, packaged, promoted, and sold. An in-depth understanding of what people want and need and what they like or dislike should come first. Edison thought deep understanding of human beings is the fundamental basis for creativity, innovation and invention. To understand people, he used direct observation, interviews (talking with them) and immersion. He also adopted prototyping (show them what they want), which is nowadays similar to empathy stage of design thinking.

Edison's greatest invention is the research and development laboratory as well as the experimental laboratory. He knew better than anyone else that learning from failure is the basis of R&D. Edison is better-known than any extraordinary scientists because he welcomed failure. He thought *failure is father of success*. The learning cannot be achieved without failure. Learning by failure, iteration of trials, and prototyping and testing, which Edison has developed, became the basic stages of today's design thinking.

Furthermore, he was an all-round player with a keen business sense rather than a scientist specialized in a narrow field. In his Menlo Park Lab in New Jersey, a team of talented and adept makers, spontaneous thinkers, and brilliant experimenters worked together. The members included Ford the Senior, who founded Ford Motor and Tesla from Europe. In fact, he was the one who created and practiced a team-based approach to creativity and innovation, breaking the stereotype of a *lone genius inventor*.

Edison's biographer describes this intriguing and diverse team members as having a pleasant friendship, but the team's research process was a series of constant trial and error as Edison's famous definition of genius puts it, 1% of sparkling ideas and 99% sweat, is what distinguishes genius from ordinary people. His method was not only to prove existing hypotheses of experimentation but also to allow experimenters to acquire new knowledge and information through repeated failures, iteration.

Creativity and Innovation is hard work. Edison, however, believes that this innovation method could be applied to the arts, technology, science, and business combined with a great insight about consumers and markets.

DT would be regarded as a follower of this Edison-style tradition. DT is a method used to meet the needs and desires of people within the scope that can create customer and market value with current technology and business strategy by mobilizing the designer's sensitivity and creativity. Therefore, just like Edison's innovation process with exhaustive efforts, creativity and innovation through DT requires tremendous effort. First, DT should start from the customers' real problems, the need of customers. It relates to the motivation and expectation of customers. This is problem identification which uses empathy as the first stage of DT. The problem should be found by human-approach. *Well defined problem is half solved* said Professor William James, the father of American psychology. Albert Einstein also emphasized the importance of problem definition as *if I have 20 days to solve an important problem, I will use 19 days to define what is the real problem and use one day to solve it.*

In today's world where most business strategies could be learned and copied easily by *Patent Thinking*, DT can make a huge contribution to the business world. Leaders of organizations see creativity and innovation as the resource of differentiation, competition and sustainability. DT could be integrated with stages of innovation processes. Corporates, universities, schools, government agencies (central and local governments), institutions, hospitals, social organizations and individuals could learn and utilized DT as a methodology of change and transformation.

13.2   Creativity for change

What is creativity? If you have same ideas as before, it is not creative. Creativity is defined generating *novel and different* ideas. In addition to that, the ideas should not be destructive but constructive. If ideas are destructive or anti-social, we usually do not regard them as creative ideas.

In order to be creative, we need to have *positive mind and attitude*. With negative mind and attitude, it is not possible to generate good ideas. From pessimistic mindset and destructive criticism, it is difficult to get creative ideas. Summarized, dream, hope, and future-oriented mindset are resource of creativity. Vision is also a good example of creativity generating resource because creativity always directs to future.

A couple of definitions of creativity can be found as follows:

1. Creativity means making change. Trying to change is the creativity (Harvard University, Professor John Kao).
2. Creativity is the behavior of thinking novel ideas. However, innovation is implementing the ideas into action (Harvard Business School, Professor Theodore Leavitt).
3. Creativity is thinking about future and preparing for it. Thinking about past and/or present is far from creativity (CCL, Greensboro).

13.3   Wake up call for sleeping giant "Creativity" within us

Everybody is born creative. Children under the age of 5–6 years are all creative. They have curiosity and always raise questions. Kids have a lot of ideas. They want to know everything. Children are asking questions endlessly. However, when they start schooling and enter kindergarten, their creativity starts to wane because teachers in the kindergarten begin to teach them what is right and wrong. Kids begin to look for right answer from this first educational institution. Primary schools, middle and high schools as well as colleges make children to be less and less creative. Their creativity fades away and they stop raising questions. In addition to that, people learn conformity, norms, and taboos of society which are mostly against creativity.

After college education, when they get job in corporate or organizations, their creativity become less as the social system requires more adaptation and conformity. When they are approaching near retirement, their creativity is the lowest level of their life-span. However, when people are retired from their jobs and finish the job career, their creativity starts to rise because they become free from corporate norms. The rise and fall of creativity for individual shows "U-shaped curve."

Everybody is born as a creative creature. As result, the competency *creativity* is always with us. However, it is like a *sleeping giant* within us. The problem is how we can wake up the sleeping giant. This huge giant needs a *wake up call*. In order to wake up our enormous potentiality *creativity*, we must have the need and motivation to be creative. Making up our mind to be creative is not sufficient enough to be creative. In addition to that, we need to know the way, method and strategy to reach the goals. As an example, if we want to go *Real North*, we need the compass which shows us the right direction.

Although we are all creative, *creativity style* would be different. Because of the difference of styles, people are in need to work collaboratively. If team members have only one same creative style, the team is no more creative. To find out what kind of style I have, an assessment can be done using creativity style test. In the test, there are four styles of creativity. Someone has only one style but others have two styles at the same time. We can find the difference between four styles as follows:

1. Explorer style: People who like to find out something new. Also prefer to solve problems which have been regarded difficult or impossible. People who like to do something new and look for entirely different information.

2. Artist style: People like to use his/her imagination. They like to find and create his/her vision. They also like to break rules and regulations.
3. Improving style: Look for positive and negative sides of a novel idea. Prefer to do step by step. Likes to approach problems without risk. Look for stability of team atmosphere. Likes to make less mistakes.
4. Experimenter (fighter/warrior): Likes to experiment and implement novel ideas. They are usually aggressive so that their ideas should be accepted and implemented by team. Encourages team members to be action-oriented. They do not worry about to be failed.

Creativity is related with future. Creativity focuses on future not past and present. People who do not think about *future*, they cannot be creative. Future thinking is connected with hope, dream and expectation. Future is created by vision and dream.

We can find out our vision and mission if we think deeply about ourselves and raise deep questions; *what do I want to be?* Parallel with this, we can also think about *What do I want to do?* Related question is *What do I want to have?* To be creative, we should know not only *what is creativity* but also *how to be creative.* Creativity requires *How to implement creative ideas* better than *nice to have good ideas.* To be creative we must have following elements in advance:

1. Knowledge and expertise in his/her specific fields.
2. Internal motivation to search for novelty.
3. Skill and technique to generate creative ideas.
4. Communication and negotiation skills
5. Team-building skills with members of the team.
6. Leadership and followership.

13.4 Creativity for post COVID-19 pandemic

The world is changing every day. We human beings are developing every week. Technologies transform every month. Organizations and societies transform every quarter. For change, development, and transformation we are in need of creativity. If we look at developmental stages of human being, it consisted of several stages: after birth, a baby childhood. After childhood, adolescence stage comes. Next, follows adulthood. When an adult is aged, getting old, the retirement stage comes. Elderly people who are aged, the aging is divided into three stages: young-old, mid-old, and old-old. As human beings, we have no way to avoid the developmental as well as change stages. Change as well as development has two faces: positive change and negative change.

During last 3 years, we human beings have been suffered from COVID-19 pandemic. It approached us without any notice in advance. It is still with us. COVID-19 has impacted every inch of human lives. It forced us to change our habits and behaviors which human beings have learned since many million years ago.

The typical new pattern of mandatory behavior was, keeping the social distance with friends, colleagues, and people. However, this new mandate is totally against human nature. People feel safe and comfortable if they meet together, eat together, and work together. However, the pandemic forced us not to meet together, not to eat together and not to work together. Instead we must rely upon Zoom or Skype, the electronic devices.

The "social distance" has changed the relationship of couple, friends, colleagues as well as social lives. During pandemic, which is still on-going, people worked far from their workplaces. Returning to the office amid the seemingly never-ending pandemic is the most complex challenge. Corporates are facing this challenge collectively and worry about if workers could come back to their office again.

Learning new ways of thinking and behavior during and after pandemic is also challenge which workers and organizations are facing collectively. Pandemic made stress for workers. How to find resilience for workers is the challenge of corporates. Before the pandemic, commuting between home and office was routine and everyday life for most workers. However, because of COVID-19, RTO (Returning to Office) became the hottest issue for most companies. Is it possible that the workers returning back to office because workers learned RTO might be dangerous? How would corporates keep the productivity if workers are staying at home and keep working? How can the efficiency and effectiveness of work would be guaranteed as before although people are working remotely? The emergence of this vague and expansive new breed of pandemic speaks to the desperation of employers trying to sort out whether, how, and when they and their workers should return to traditional offices. Returning to office without danger is one of the most important challenges which needs creativity. Also, it needs solution.

13.5   Culture and creativity

Creativity of children is heavily dependent upon the *ways of child-rearing*. How the parents raise their children from the early developmental stage plays an important role for the development of creative capability of them. If we compare Eastern culture to Western culture, we can find remarkable difference between two different cultures. Parents in Western culture give relatively more freedom to children so that they can decide by themselves from the early age. In comparison, in Eastern culture, parents have usually more influence and power to help their kids' decisions.

Developmental psychologists report that creative competency of children depends much on the ways of raising method. I have observed and compared the attitude and behavior of parents when they interact with kids with younger age between Korean and German cultures. Korean parents advise and intervene whenever kids should make decision. For example, when the families go shopping to buy a toy for kid, mother plays an important and decisive role than kids themselves. In compared to that, German parents give more decision power to kids to choose toy. Entering schools and college is always hot issues between parents and children in Korea. In addition to that choosing college and majoring field, the influence of parents become much bigger. However, the developmental psychology studies show that if children have more freedom to decide by themselves, their capability of creativity become bigger than dependent children.

Center for Creative Leadership (CCL) in Greensboro published a creativity checklist for children-rearing. It focuses on the relationship and attitude between parents and children. You may check following items to find out if you are raising your children creatively:

1. In general, parents respect their children.
2. Parents give much freedom to children when they have something to decide from early age stage.
3. Parents expect their children would behave with responsibility as well as independence.
4. Parents set limitation of freedom for the children.
5. Parents do not interfere with kids' decision making.
6. Kids easily find adults around him/her so that they can benchmark and learn from them.
7. Parents let children can distinguish between what is right and what is wrong.
8. Parents like to go out with kids when they go shopping (or travel, visit relatives, entertainment parks, etc.).
9. Parents answer the questions of kids with delight.
10. Parents let children decide if they go to buy toys.

13.6   Creativity to generate novel ideas

Idea generation is a kind of act and habit. Habit requires repetition. If we do same things again and again, it comes automatically as a habit. Eating is a kind of habit. Sleeping is also habit. If you practice to generate ideas daily, the numbers of ideas increase. Let us have an exercise to generate ideas. This is an exercise for idea generation:

> You know traditional Coca Cola bottle. It looks like the figure of a woman and it was very popular for 50s and 60s. After we drink Coke, the bottle is empty. Here is the question: Please think of what purpose we can use the empty bottle. You have three minutes to generate ideas. You should write ideas on the paper as many as possible. Now, please start! Time is only three minutes!

The second exercise of idea generation is *vinyl umbrella*. We use vinyl umbrella when it is raining. The instruction is:

> How can we use vinyl umbrella with wood besides protect us from raining? Please write down on the sheet of paper as many ideas as possible within three minutes.

The result of both exercise is interesting. At the beginning, people would write only a couple of ideas to use empty bottle or vinyl umbrella. If this kind of exercise is repeated, they get more ideas! Repetition and iteration in DT make people master of idea generation.

People get ideas whenever they encounter following situations:

1. You would like to improve and change the inconvenience which surrounds yourself and your activities.
2. You have positive attitude generating lots of novel ideas
3. Trying repeatedly to find out novel, fresh, and attractive solution whenever you face problems makes you an expert of idea generation.
4. You have a habit to write down your ideas whenever they pop up in your mind. Even small and invaluable ideas at the beginning should be written down in your notebook.
5. Try to find out anything that you regard as problems, and find out the solution.

When comparing with above mental set of idea generation, we can also include following actions so as to enter into the kingdom of idea generation:

1. We have the ability to think about fresh and novel ideas in order to make our living and working environment better and comfortable.
2. We generate good to save time, get rid of unnecessary cost and inconvenience in our personal and work life.
3. Excellent ideas could add value and make impossible to possible.
4. In order to be idea generator, you should have a habit to think about interesting, meaningful, and efficient ideas as a routine.

13.7   Idea racing system: how to keep ideas

Human being is the only animal which can think and generate ideas endlessly. Other animals may also think, however, they cannot do it like human beings. In general, we are the only animal on the globe that can think and generate ideas. We produce ca. a couple of thousand ideas a day. However, we forget most of them every day.

Since we make so many ideas per day, sure there might be some good, useful and productive ideas among them. Most of them may be not so much valuable. However, we forget most of them. We usually forget 99% of generated ideas. Psychological studies show that human beings have the ability of retention at most 5–7 items at a time. This means that if we have more than five or seven number of ideas at the same time, we would forget them. In order to keep the ideas, we need either to write or to record them so that we can memorize them.

Forgetting curve is opposite to retention curve. As we forget easily, we need to find out the devices and technology which would supplement and support our retention capability. One of them is as we know very well "writing." If we have written down our ideas on a sheet of paper or notebook, we can read it later and do not forget. The other way is recording. As we all know, tape recorder helps us so that we can listen the idea later and do not forget. Electronic device like smartphone would fulfill the role of writing as well as recording at the same time. It also can be used to photograph the ideas written in the paper. To keep ideas, of course, we can not only write them down but also record them using various analogue and digital devices.

Traditionally, the easiest and cost-effective way of keeping ideas is to write them in a notebook whenever ideas pop up in our mind. In order to keep and not forget our valuable ideas, we need to prepare a notebook and a ball pen. We write ideas in one notebook. If you decide to keep your valuable ideas, you should prepare one notebook which you can use for a couple of months. It is vitally important that you use one notebook so that ideas would show sequentially. In case you keep your ideas in the different or separate sheets of paper or a couple of different notebooks, your valuable thinking would be not collected but easily forgotten.

If you like to start the Idea Racing System (IRS), you should prepare following items:

1. A notebook which is flexible. You can fill in the sheets and pull out them.
2. The notebook size is important. The A5 size is good because it is as big as you face size. You can read the ideas of notebook with at a glance.
3. Use ball-pens which does not permit to be erased easily.
4. It helps a lot if you use a couple of colored ball-pens. They allow you to distinguish the ideas.

Writing has been important method for counseling and therapy in clinical and counseling psychology. Stress, neurotic issues, and psychotic problems of patients should be either recorded or written down on the paper with agreement of patients. With the written contents of psychological problems, both clients and clinical psychologists could interact efficiently to solve problems.

Normal persons are not exception. If you have something to worry or stressful issues so that you could not get in sleep well, you may write them down on a sheet of paper. After you have written them, you may get in sleep more easily. Because if we write something, we can reflect it and observe ourselves objectively. Also we can observe ourselves with the problems from different perspectives.

Keeping ideas is a challenge. Normally, most of us forget valuable ideas because we do not use retention supporting devices. Let me introduce extreme situations which are related with generated ideas. What would you like to do to keep your valuable ideas if you are facing following situation?

1. You are taking a walk in the forest. Suddenly, a very good idea for a solution of your problem pop up! By chance you have neither ball pen nor paper. You would like to keep this idea until you reach home. You forget to bring smartphone with you to record ideas. What would you like to do in order to keep your valuable ideas so as not to forget?
2. You are taking shower now. During your refreshment, a couple of good ideas related to your work popped up. There is nothing to write or record in the shower room. What would you like to do so as to keep these ideas?

3.   You are now swimming in the pool. During swimming, you get an excellent idea for your problem. In order to keep this valuable idea, what would you like to do?

These are extreme cases. However, we face this kind of dilemma nearly every day. Most people are reluctant to write something down. They do not bring writing material or recording devices with them. However, the biggest barrier of retention of ideas is people rely upon their memory ability too much. In order to solve this dilemma, we suggest a new method "Idea Racing System" (IRS). IRS is a technology that we do not forget ideas in our daily lives. Ideas are very easy to be lost because we do not have cognitive retention capability more than seven ideas at a time. Idea-racing system suggests that we should write down ideas in a notebook according to the calendar date, serial numbers of total ideas, and the number of ideas on that day using the same notebook until it is filled.

MZ generation is grown up with smartphone, computer, and IT devices. The old generation who are not grown up with these digital devices are rather familiar with writing habits. Both MZ generation and old generation would find IRS helpful so that they could keep their ideas better and longer learning the IRS.

13.7.1   Method of IRS

IRS consists of several principles and rules so that using IRS would be a habit. IRS is regarded as biological extension of brain and physical arms. When compared with ISR, smartphone is more technology-oriented, which we regard as inhumane. Let us look at the principles of IRS to be a useful technology:

1.   Keep using same size notebook until it is filled.
2.   Think and generate new ideas daily. At the beginning 3 ideas, later 50 ideas.
3.   Write these ideas into your notebook. Do it by yourself.
4.   Draw pictures as many as you can. Pictures speak louder than letters.
5.   Talk to your neighbors or colleagues about your ideas. This is a kind of evaluation process. You can do it alone and/or with colleagues.
6.   Implement the best ideas as soon as possible. Prioritize your ideas frequently and decide when, where, and how to implement.

Drawing pictures instead of writing sentences is applicable to kindergarten kids who did not learned writing. In the history, Leonardo da Vinci and Thomas Edison were inventors who used drawing to keep their ideas instead of writing them. Usually drawing talks louder than sentences. Leonard da Vinci drew the prototype of flying about 500 years before. Thomas Edison, his famous Laboratory in New Jersey, is still full of inventive drawings which are not yet opened to public.

13.7.2   IRS: its basic principles

IRS has several advantages. Complex, expensive and difficult method is hard to be adopted. However, ISR is not complicated but easy to learn. In comparison to IRS, TRIZ (teoriya resheniya izobretatelskikh zadach, inventive problem solving) developed by Russian inventor and writer Genrich Altshuller [76] is very difficult to learn. It requires many rules and regulations to implement innovation. Simplicity of creative method offers access to average persons. Everybody can understand IRS easily. Many people can learn IRS at the same time. Let us check the advantage of IRS as follows:

1. Easy to do. Write ideas whenever ideas pop up.
2. Simple to implement: a A5 notebook and ball-pen are enough.
3. Everybody can learn, from kids to adults, and the aged can also learn it.

Kindergarten kids with the age 3–5 years old can learn and use IRS. If the kids cannot read and write the language yet, they may use drawing ability because everybody can draw if they were given crayon and a piece of paper. At the beginning, kids learn drawing by observing plastic toy animals. They observe animals and find out the uniqueness of toy animal. Observation increases the creative ability of kids.

IRS costs nearly nothing so it is economical. We can bring IRS notebook wherever we go. If we bring IRS notebook with us, we can generate ideas whenever necessary. There as some rules which help us to continue using IRS:

1. A5 size notebook (6×8 inches) is best.
2. Do not use a couple of separate notebooks. Keep only one notebook until it is filled out.
3. Write ideas whenever they pop up. Bring your notebook wherever you go.
4. Do not skip pages. Keep the sequence without jumping pages.
5. Do not make sections in a notebook.
6. Write date, number of pages, and number of ideas per day.

Let us find out why we should write ideas in the notebook using IRS. These are the logics why we have to use IRS to keep our ideas:

1. If you do not write your ideas, you forget most of them easily.
2. Notebooks are backbones of your daily life and work. You can use not only at home for personal purpose but also in your work. Or you can use two separate notebooks if you like.
3. Notebooks act as an extension of your brain.
4. Ideas could be reviewed and chosen in order to be implemented. From time to time, we should evaluate the ideas in order to find important and urgent problems.
5. The number of ideas could be increased if you continue using ISR. At the beginning, you might write only one idea a day, then three ideas a day. If you like you can increase 50 ideas a day.
6. Brain could be innovated through the habit of thinking and writing.

13.7.3   Application and magic of IRS

IRS can be applied in many areas. For example, it is used by researchers and engineers in R&D centers who are in need of many ideas in order to develop new product and devices. Corporates and startups which develop software, new business and new concepts of design are adopting IRS. Researchers in R&D centers, university students and professors, businessmen and salesmen are using IRS.

Recently, a large university adopted ISR for the whole freshman students. Nearly one thousand freshman students started to learn IRS. During academic orientation, students took part in the IRS lecture. University offered them A5 size standard IRS notebooks. IRS consultants delivered a series lectures on IRS. Freshmen could start writing IRS notebook from the first academic day.

Students were divided into different groups. They became a member of a team, so that they could work collaboratively. Assignments for idea generation were given both off-line and on-line. Students have access to ask questions to consultants. They could communicate with IRS consultant by an e-mail if it is necessary. Feedback was given by computer. After 3 years of IRS practice, nearly 20% students of this university was using IRS technology.

IRS is a remarkable method to apply our personal life. Since the notebook does not need to be shown other people, it is 100% private!. It is personal space and domain. Vision creation and mission statement could be done via IRS. Personal dream and future creation would also be objectives of IRS application. These private and personal issues should be in the notebook and could be amended several times if necessary. Life design, design of future, search for meaning of life and life purpose statements are also frequently included. Poem writing, thesis study and research design are also objects of IRS.

If we use IRS technology, we can get three kinds of creativity:

1. Serendipity: getting excellent ideas by chance
2. Ideas by on-going thinking
3. Ideas on the spot request

The biggest challenge of IRS is how we can keep IRS writing as a habit. It should be a part of person's habits like eating and sleeping. During our practice of IRS consulting, we found the 3.3.6 rule. First 3 days are very difficult to continue idea writing in the notebook. Three weeks is a kind of small mountain which beginners should climb. It is called critical point. After 3 weeks, 3 months would be another high mountain to climb. If IRS writing is continued more than 6 months, it is a sign of habit. Summarized, the critical points of IRS are:

1. 3 days: Survival point.
2. 3 weeks: Critical point of IRS.
3. 3 months: Idea writing starts to be habit.
4. 6 months: Probability to keep IRS writing is bigger than before.

13.7.4   Collecting dreams for psychoanalysis by IRS

If you want to be trained as a psychoanalyst you have to collect your dreams. The collected dreams of candidates should be analyzed by senior analysts. This was a part of obligation who want to be qualified as a psychoanalyst in Freudian Psychoanalysis. The challenge is how to collect the content of dreams. During sleep, we often dream of dreams. The length of dream is very short. It could be no more than a couple of seconds although it looks like a long story. In the morning when we wake up, we recognize that we had dream during sleeping; however, it is difficult to remember.

Some part of dream we can recognize, however, the whole story is nearly impossible to remember later.

If you want to get the whole story of dream in the morning, you must write it down on a sheet of paper when you dream a dream during sleep. This means that you have to stop sleeping and get up and write it. It is like torture to break sleeping and wake up and write the content of dream. The best way to collect dream is like this; stop sleeping and get up and sit down on the table. Write down the dream which you have dreamed. With this method, you can collect your dream but the sweet sleeping is gone away!

From the same context, IRS requires to write down your ideas as soon as possible. If you do not carry your IRS notebook with you, jot down the ideas on the separate paper and later you can write in the notebook.

13.7.5 Time and places for getting ideas

If you have something to think and want to get idea, where would you like to go? Earlier people used to go library or bookstore to find reference books. Also, people go to downtown and they observe other people or show windows of shops. Looking around his/her surrounding to find out information and knowledge related to the problems is also a good method to get ideas. Nowadays, people start Googling or Internet surfing. Young people like to go Starbucks with notebook computer instead of library. They connect Internet and start to navigate information. Many people use smartphone to search information also. As we have seen, the landscape of idea-searching became very different from earlier days.

According to cultures, people prefer different places to think and generate ideas. In a cross-cultural survey researchers asked *when do you feel comfortable to get ideas?* The Americans answered they like to generate ideas (i) during driving (44%), (ii) while taking a walk (35%), and (iii) in bed, before getting asleep (29%).

In compare to that, Japanese get ideas (i) in bed, before sleeping (52%), (ii) while taking a walk (46%), and (iii) in the car (45%).

The other common places for idea generation to both cultures were on the table at home, at the coffee shop or bar, at the toilet, public bath or taking shower, and outside the workplace.

Idea generation has deep relationship with right timing. During a day, when people like to think and generate ideas were:

1. In the evening (46%).
2. Late at night (44%).
3. In the bed before sleeping (34%).
4. In the afternoon (23%).
5. During day (18%).

13.7.6 Creativity and illogical thinking

Bertrand Russell said that "It is a shame – Stupid people are always sure about what they are doing, Only the bright ones constantly worry about their BELIEVES and what they are DOING." He insisted that doubt as well as worry about knowledge is sometimes necessary. From this logic, curiosity, which motivates and forces us to know, is the mother of questioning. If we are in need of knowledge and information, we ask questions. Kids are constantly asking questions. They are full of curiosity. However, when kids grow up, they stop questioning. Adults ask less. People who are sure about something, they stop asking questions. Curiosity has intimate relationship with creativity because asking questions is the starting point of curiosity.

We have discussed "Creativity is generating novel and useful ideas" whereas "Innovation is implementing ideas and making things and product as a reality." We need to bring creative ideas into innovation, if not they are useless. Creativity has to do with change and transformation. The following four categories should be thought if change would take place. They are nature, society, thinking and technology. These include evolutionary change of butterfly, the growth of cities, personal thinking and development of technology.

1. Nature: egg – caterpillar – cocoon – butterfly.
2. Society: village – police – city – metropolis.
3. Thinking: phantasm – imagination – creativity – innovation.
4. Technology: gliding – propeller – supersonic – spacecraft.

Creativity and innovation are difficult because they are not simply matter of technology. It is matter of cognition and thinking. Logical innovation brings only small change but illogical innovation brings big change. Sustainability of a corporate is request the biggest challenge. If we follow only logical innovation, it is not sufficient. It should come from illogical thinking. Ideas from illogic can challenge big problems which keep the corporate sustainable.

In order to follow illogical thinking (widerspruchsorientierte Innovation), we have to look at the historical development of contradiction-oriented philosophy:

1. Parmenides (515–450 BC). Unity – polarity of the world: material – nous, motion – rest, arrangement – chaos, arise – decay.
2. Heraklit (540–480 BC). Logos: All things are in the motion. Fight is the father of all things.
3. Socrates (470–399 BC). Maieutics: Create knowledge and experience by asking questions.
4. Platon (427–347 BC). Dialectic: Thinking in categories of being. Essentials despite ostensible.
5. Aristotle (384–322 BC). Formal logic/syllogism. Dialectic as an instrument for verifying thesis.
6. Galilee (1564–1642). Emporiums: Systematic experiments despite of dogmatisms.
7. Hegel (1770–1832). Idealistic dialectic: Thinking the thought of good prior to creation.
8. Marx (1818–1883). Materialistic dialect: Laws of development of human society.
9. Nietzsche (1844–1900). The permanent occurrence.   ·
10. Altshuller (1926–1998). Method to inventively overcoming target conflicts (TRIZ).

11. Coware: innovation teams.
12. DT: out of box innovation.

13.8 DT as a vehicle of creativity

DT appeared in the literature since a couple of decades ago. It has been rapidly expanded in different disciplines in a couple of advanced nations. Why this creativity – and innovation method has attracted attention of so many people? DT has uniqueness and distinction in comparison with existing methods of creativity:

1. DT process is relatively simple. It is easy to understand at the beginning. Most DT approaches are organized in five stages targeting empathy, problem statement, idea generation, prototyping and testing. However, there are 3-stage and 6-stage versions of DT.
2. It is not only personal but also group focused. If the members of group use DT tools together, it is fun and easy to solve problems (mostly small problems at the beginning).
3. It deals with not only small problems but also tricky problems that are difficult to solve.
4. Tricky and difficult problems would include following issues:
   (a) Low birth rate in many countries.
   (b) Defense industry. The number of male recruiters in Singapore military service is decreasing. The female recruiters are increasing instead. The military equipment and weapons for male soldier are not suitable to females.
   (c) Wicked problem, such as the transportation system in a metropolitan city.
   (d) Taxation issues.
   (e) Support and assistance problems for the elderly.
   (f) DT is culture-neutral. Team members in different countries conduct problem-solving sessions inter-culturally.

DT is based on the basic human needs and the psychological motivation of human beings. For example, creative confidence, resilience, and communication of members are backbone of DT. Because DT is human-based technology, it can be transferrable from culture to culture.

13.8.1 Why is DT attracting attention?

Why DT became a global trend? The reason is that not only engineers and scientists but also white-collar workers engaged in intellectual work are also in need of creative ability in order to solve problems. And this ability could be learned through DT. This became the driving force of DT in many knowledge societies.

DT became popular because the existing problem-solving and invention methods like TRIZ are difficult to learn. It is mainly useful for engineers. It is logic based and left-brain style approach. Of course, DT does not work well for all kind of problems. The problems that DT hits better are *wicked problems*. It emphasizes right problem definition using empathy approach. A poorly defined problem is hard to solve. DT enables us to define the problems better than any other approaches. Well-defined problem is the output of human approach. It means that we should find out the problem from the owner of it. In order to define problem, Design Thinkers use Observation, Ask, and Immersion. Observation means look at the problems, Ask means talking and interviewing, Immersion means to be as a real problem owner.

If we face wicked problems, unknown problems, and tricky problems, DT would be one of the best solutions to be applied.

There are three keywords that run through innovation: inspiration, idea and execution. DT is appropriate tool for projects that want to be completed creatively focusing on inspiration, ideas and execution. *What if we all work together?* The slogan of IDEO, a DT-based consulting company that is beyond any individual uses team-based DT process in which a large number of small teams collaborate effectively. In addition, the project rooms of DT of IDEO are the learning and making space. Also the learning space of Stanford University's d.School, HPI Institute at Potsdam University and the innovation space named Gym of Procter and Gamble (P&G) are all contributing to create innovative culture of organizations.

The new DT space during and after pandemic has been dramatically transformed using robotics, electronic white board, and remote communication. HPI Institute of Design Thinking at Potsdam University has been digitally transformed; however, human-based atmosphere for creativity is secured.

The basic approach of design thinking acts as a means and driving force to achieve innovative results. World-class design schools in Europe and the United States, MBA courses of prestigious universities, design companies, and global innovators and leaders are analyzing the global trend that applies DT for the creation of sustainability of their organizations.

Companies in a mature society face difficult situation if they do not keep innovation continuously. Companies in advanced countries need to change their profit model because competition from other large companies and SMEs in emerging countries intensifies. "My job is disappearing right now. Therefore, if we do not create a new business, we cannot survive." is a saying that managers of companies speak to the population of workers.

Fierce competition is taking place between white-collar workers in developed countries and cheap, high-quality workers in emerging countries. As a result, wages for workers in developed countries are falling (Indian engineers working in Silicon Valley). The way to deal with this challenge is to create a new value so that sustainability could be kept. This value creation expands from technology transformation to humanities, culture, art and entertainment.

13.8.2 DT for value-driven innovation

The world has been changed. The speed of change is faster than ever. Every corner of our lives is impacted by digitalization, big data, cloud, and robotics. Summarized, we have to find changes of our daily lives as follows:

1. The advent of the Internet, smartphones, cloud computing, Artificial Intelligence (AI), social networks, etc. have brought changes into the social structure. As a

result, in the past, creative and expressive activities were the exclusive property of some people, but today, everyone should become a knowledge creator.

2. To be specific, Facebook has more than 1.3 billion users worldwide. More than 300 million people are using the video sharing social media called Telegram. Taking pictures with a smartphone and sharing it with friends through Instagram or TikTok are routine activities of young people. Communication by SNS as well as processing the picture into nicer picture have become everyday entertainment of young people in many countries.

3. The parody videos posted on YouTube becomes popular. The number of jobs like websites design, operating social media communities, and marketing by SNS are increasing.

4. 3D printers and laser CNC (machine tools) could be used easily. Desktop manufacturing or maker movement becomes a commonplace. So anyone can design and create anything necessary for their own lives (DIY: Do It Yourself). As a result, one-man manufacturing era has arrived.

5. Designing daily life by using creativity and applying digital device for commercial purpose, these activities become either side job or main job.

13.8.3 Combining machine and human work

AI enabled us to do machine learning and deep learning. As learning using AI is developing, innovation could be collaborated easily with DT which is strongly human-centered. It is expected that machines would replace human beings. As can be seen in the Go match between AlphaGo and Go Master Lee Sedol, an AI Go program proved the role of AI by big data. The deep learning would grow rapidly. However, AI also has some dark side of it.

Oxford University scholars Frey and Osborne showed in his study *The Future of Employment* [77] about jobs that will disappear as computers evolve. He also listed jobs and specializations that will continue to survive. They fall into the following two categories as jobs that are difficult to replace. These are occupations related to deep communication with people, such as healthcare, learning, and psychology, and those requiring creative power such as design, engineering, and management.

As AI develops, AI can provide more efficient answers in data analysis and pattern understanding, so work centered on analysis is classified as a field with high substitutability. Nevertheless, it is natural for humans to provide solutions and make decisions about the contents analyzed by the machine.

13.8.4 The advent of the age of self-esteem and personal happiness

Self-satisfaction is a way that individual finds happiness in his/her own life. In mature societies, the values pursued by people are gradually changing from material satisfaction to mental and psychological satisfaction. By the 2050s, the world population is expected to reach 9 billion people. Therefore, it is the task of mankind to divide the limited resources among the number of people far greater than it is now. As a result, the twenty-first century will be "the age of knowing how to be self-sufficient."

Creativity on a daily basis increases your sense of well-being. Individuals will increase their sense of happiness if they design their own life style and live a personal life.

As an outcome, *Design your Life* became popular to people who want to live meaningful, productive and satisfactory lives. Students at college use this DT approach to create his/her future. They want to find out what they want to do, to have and to be. Middle aged people also enter DT courses in order to reflect their lives or to find different course of life which gives them meaning of life.

Positive psychology, meditation, reflection, scenario planning, and psychological safety combined with DT enable people to design their lives.

It offers opportunity to contemplate and reflect personal behaviors in his/her corporate. Contemplation offers value seeking, mission and purpose statement. The renewal of personal lives through *Design your Life* gives opportunity to change personal habit, creativity, and lifestyle. Life design offers a field and well as culture focusing on vulnerability, authenticity, and feedback while psychological safety is secured. Middle aged people who are burnt out by current jobs or unhappy with current family situation also volunteer *Life Design* courses to find vitalization. Life design by DT are combined with emotional intelligence, self-awareness, purpose statement and creative habit.

13.9   Misunderstanding of DT

When the words DT are used, people often associate it with a specific department of college. They think of department of arts or courses of designers. The word "design" is frequently associated with image of fashion and art. As a result, people say that DT is necessary only for those who study fine art, and it has nothing to do with me. Therefore, people turn their backs on DT by giving reasons such as no artistic talent or no aesthetic sense.

I met a Korean woman in Singapore who graduated from a design school in Chicago. After she majored economics at a college in Korea, she worked a couple of years in a corporate. Later she entered Design School in the United States for 3 years. She is currently working in a bank in Singapore. Her job is new financial product development for customers of bank. She recalled that while she was attending at the design school, she never had any class of the so-called design-related courses. She never heard decoration or coloring which fashion designers are doing. After she completed 13 different projects as a team member, then came her graduation.

DT is a method to find solution of problems. To solve problems, we need creativity. Problem can be solved more effectively as a team than individual. If we have a problem, we apt to consult with expert in order to get advice or solution. This solution from an expert is mostly not the right answer because the expert does not know what is real problem of the client. In many cases, a client who has problem does not know what is their actual problem. This is the reason why "problem finding" is most important. William James, a professor of psychology at Harvard University, who was called the father of American psychology, once argued that "a well-defined problem is already half-solved." This phrase is also frequently quoted by Steve Jobs, the founder of Apple. Albert Einstein, who won the Nobel Prize, said, "If I had 20 days to solve a problem, at the beginning I would spend 19 days defining the problem, and the rest of the day I would use to solve the problem." The advice of these two scholars remind us how important it is to define a problem correctly. We do not spend enough time to find out and define an actual problem. In fact, we are often surprised to learn how often we waste our time solving wrong problems that are not defined correct.

DT is neither self-centered thinking nor expert-centered thinking. It is customer-centered thinking. In other words, this is a user- and consumer-centered thinking. Problems should come from the problem owners. Therefore, DT is human-centered problem-solving method. The five stages of design thinking process could be applied to many fields and organizations (company, university, hospital, government institution, etc.) However, it is sometimes important and necessary to adjust and modify the DT according to the organizational culture, countries, and industries. Furthermore, the DT process is iterative, enabling exploration of insights into problems and issues. Companies emphasizing the importance of customer experience are using DT. These are, for example, automobile and smart phone producing companies. They advertise we do not sell to customers a vehicle or smartphone but a new experience.

13.10 Reflection of DT experience

COVID-19 has impacted the expansion of DT because of the mandate of "Social distance." However, DT have been taught extensively. Universities teaching DT to students as regular/irregular courses are Stanford, Harvard, UC Berkley, Toronto, Virginia, Texas State University in North America. School of Business at Toronto opened DT as a regular course. Darden business school at Virginia University is active teaching DT.

In Europe, the HPI Institute, School of Design Thinking, Potsdam University offers courses not only students but also to managers from corporations. North Europe is not an exception. The NTNU, the National University in Trondheim, Norway is active to deliver DT to students with various majoring fields. They learn creativity, teamwork and product development and product design. Bergen University in Norway offered 9 months Executive DT Course for enterprise managers and government employee. Three different Universities in Sweden operate a DT center collaboratively. Aalto University in Finland is famous for teaching DT not only domestically but also internationally.

Japan is active expanding DT. Keio University, Tokyo University and Kyushu Universities built DT consortium. Chinese Universities are collaborating with Western universities. Singapore is very active. At the beginning, one of the leading Universities in Singapore contracted with University of Toronto and dispatched professors to be trained. After a couple of years, they teach and consult DT not only in the colleges but also corporates and government organizations. DT is applied in many fields of Singapore societies. Healthcare, education, and consulting areas are some examples. DT experts are consulting for hospital, library, transportation, tax collection agency

IRA, defense industry and financial institutions. The government is supportive to use DT so as to make the lives of citizens happy.

I have introduced DT in Korea about 12 years ago. Colleges and universities were enthusiastic to adopt DT in order to teach creativity and innovation. Engineering colleges adopted DT actively. DT became obligatory course in some colleges. Students from different majoring fields volunteered to take part in DT workshops and learned the process of DT as solutions for problems. They learn not only creativity but also collaboration and leadership through DT. Medical schools were active to learn DT in order to offer positive *Patient Experience*. Also, hospitals took part in the new trend. It is related to patient experience which ultimately decide the sustainability of hospitals.

I would like to introduce a case which our consulting team have delivered. US national agency "Hawaii State Tourism Organization" organized a visiting team and visited Seoul. About 30 tourism-related experts in Hawaii visited Korea as part of the travel revitalization project of Hawaii. They wanted to promote the visit and travel of Koreans to Hawaii. The visiting team asked a DT workshop to us in order to develop attractive Hawaii tour program. Their problem was how to promote Korean travelers to travel Hawaii. And also they wanted to know what kind of travel experience do the Korean travelers expect. Hawaiian delegates were managers from Hawaii-based hotels, car rental companies, travel insurance companies, and museums and art galleries, and airlines.

In addition to the 30 tourism experts from Hawaii, 30 Korean tourism managers joined the visit. Thus, with 60 participants from different cultures, genders, and ages, the DT workshop has been conducted. It was 2 days course. The highlight of this course was that travel industry officials from different cultures were dispatched to Incheon International Airport, where travelers were entering and departing. They have as a team observed, interviewed and felt like travelers who were coming inbound or outbound.

Participants from two different cultures presented their results at the end of workshop, which was very impressive. A couple of teams presented that "we were surprised that we know so little about the travelers' sentiment although we have worked in this industry such a long time." Other teams confessed "we have developed travel product without thinking about traveler's expectation." Also, one team said "we have never thought how bad our travel service was to the visitors."

These insights were product of problem finding "Empathy Stage" of DT. Also talking with travelers who were coming back from their travel gave unexpected ideas about travel experience. Three approaches of empathy stage, observation, interviewing and immersion gave a lot of impact for the experts of travel business which they have not known for such a long time. They realized and confessed that they knew first time in their career what the potential expectation of temporary customers. Some confessed that they knew nearly nothing and developed and sold travel products to the customers.

Chapter 14

Conclusion

Manuel J. Fernández Iglesias[1] and Íñigo Cuiñas[1]

This chapter summarizes the main ideas discussed throughout the book. For this, we carry out an overarching critical analysis about the content introduced in the previous chapters, identify remaining open questions, and foresee the evolution of this methodology and its role in the education of future graduate students.

This book addresses the design thinking (DT) methodology as a basically sequential process that is developed through the five phases of empathy, definition, ideation, prototyping and testing. DT is a design process based on the fact that solutions to problems must take into account the actual people who experience those problems. We do not cater to the needs of archetypical or standardized users and we do not base our solutions on an abstraction of a collective or on a generalization that ignores the specific characteristics, even the expectations or aspirations, of the people who rely on us to help them overcome their challenges.

Through empathy, we try to capture all the relevant information about the problem and the people who deal with it. We become the user and try to be part of their day-to-day lives in order to understand their motivations and hopes. We become aware of the technological context in which we operate, as well as the ethical and human considerations applicable to our challenge.

Once all the relevant information is captured, we study and classify it to distill the information that helps us to clearly identify the challenge we are going to face. We select and fully characterize the specific problem that we are going to solve. In general, we will seek a solution to the challenge that will have the greatest impact on the people for whom we work, taking into account the technical and human context. We express this problem as a point of view, as a motto that motivates us to act and guides our solution-finding process.

The point of view will be the starting point to initiate a process of idea generation. We give free rein to our imagination to try to identify all possible approaches that will lead us to a solution. We then study and classify all the candidate ideas in order to identify the one that will have the greatest impact, the one that will be a viable solution to the problem initially posed that also takes into account the boundary conditions and users' expectations.

[1]atlanTTic, Universidade de Vigo [GID DESIRE], Spain

Once we have identified a solution, we share it with the users to find out whether we have really managed to overcome the challenge that we have set ourselves, that is, if the proposed solution actually solves the problem in a way that is acceptable to the people who experience it. To do this, we build a prototype with which users can interact. We share the prototype with them and observe how they interact, what questions they ask, if we observe any behavior not initially foreseen or if the solution omits any important aspect of the original problem.

Although this process is introduced in this book as a linear succession of stages, these five stages do not always occur in the aforementioned strict order. Some stages may be developed in parallel, or a previous stage may be revisited at any time to use the findings of a later stage to enhance the results of the entire project. For example, when testing our prototype with real people, we may discover that the interactions with it are not as expected, which may lead us to revise the proposed solution to take into account this mismatch between our expectations and those of the users. It may even be the case that we have not identified the problem correctly, which will lead us to revise our point of view. Consequently, we will have to look for new solutions to be prototyped and shared with the users.

Similarly, DT also adapts to an iterative or sprint-based design process. We can consider a large DT project as a succession of projects, in which we build the final solution from partial solutions or less refined solutions that evolve to the final solution. In this case, each DT project would take as its starting point the result (i.e., the tested prototype) of the previous iteration. Indeed, the testing phase of the present iteration becomes part of the empathy phase of the next interaction, so that we use the information captured from users and interactions with the prototype to identify a more elaborated context, closer to the context of the final solution to our problem. Then, we will take into account the new information captured and all the information from the previous stages to define the objectives and the specific problem to be addressed in the new stage.

In short, DT should not be understood as an inflexible and rigid methodology for solving problems. If this occurred, it would become just another traditional methodology. The sequencing in stages should be used as a guide that indicates the natural evolution of events. However, for each specific project, we can traverse these stages in a different order, trace back, perform them simultaneously or iterate them several times to broaden our vision of the problem and finally come up with the best possible solution. The information that circulates throughout the canonical stages serves to better understand our audience, the original problem, and the solutions that we are planning.

However, DT has three very powerful enemies that can compromise a project to the point of failure. These enemies are fear, resistance to change and devil's advocates. These enemies could work together to obliterate creative output through non-constructive negativity.

Fear prevents design thinkers from using their methods and tools to achieve their goals. It makes us hesitant and distracted, and it generates doubts about our knowledge and skills, compromises our self-esteem, and limits our ability to make timely decisions. To overcome fear, it is first necessary to determine what the real problem is and to explain it clearly and concisely. This explanation must be based on actual facts and must avoid any emotional element. It is very important to know how to differentiate between our expectations and hypotheses and what actually happens. Clearly separating these aspects will allow us to clarify the real problem.

On the other hand, if we omit information or opinions for fear of conflict, we are placing a barrier between ourselves and the rest of the team members, preventing the relationship with other design thinkers from being authentic and thus compromising the outcome of group tasks such as identifying a point of view or generating ideas. Being aware of the importance of honestly sharing our concerns and of listening to what the other has to say, it becomes easier to solve or clarify the reason for conflicts or misunderstandings.

Resistance, on the other hand, hinders the DT process by contaminating our goals with all sorts of things that need to be done first. It moves the focus from the search for solutions to anything other than that search for solutions. Something more urgent always comes up when we have to face a task that can move our project forward. Resistance can also come through other people, even other team members. The main trigger for procrastination is negative feelings. When you feel stressed, indecisive, overwhelmed or even bored, you are more likely to resort to this change-opposing behavior. It is a simple mechanism, based on people's natural tendency to avoid whatever is perceived as unpleasant. To overcome resistance to progress, motivation is a very powerful force, pushing you to make efforts to get what you want. In this sense, it is the opposite of procrastination because it encourages you to postpone immediate rewards in order to achieve greater rewards in the long term. Therefore, when defining goals and strategies, we must be as specific as possible. Very general or broad objectives are not motivating. They may even generate some discomfort because they are perceived as too distant or difficult to achieve. If you intend to achieve an objective that is not relevant, you will hardly feel motivated to work on it.

The devil's advocate is that person who never has anything productive to say, but immediately knows and expresses exactly why every solution initially proposed will not work. It is true that there is a positive version of the devil's advocate, personified as that team member who takes on the role of conducting a systematic critique of a course of action. The purpose of this critique is to point out weaknesses in the assumptions underlying the project, internal inconsistencies and problems that could lead to failure if a proposal is adopted. In this case, devil's advocate acts in a way similar to a good lawyer, arguing against the majority position as convincingly as possible. This role must be rotated to prevent an individual from identifying himself as critical of all solutions, moving from being perceived as an assigned role to an assumed role. However, playing the role of devil's advocate, even for a brief period, may be advantageous to the individual and to the organization.

In any case, the profile we are referring to in this case goes beyond the critical thinking and critical analysis just described. It is a role inherent in the personality of a team member, and therefore not assigned. This type of person has the ability to derail projects by shifting the focus from potential solutions to hypercritical problems that might not even be relevant after all. This person's goal is to stop any further ideas towards a solution and should be kicked out of the room.

If we overcome these enemies and implement DT in a responsible, collaborative and truly people-centered way, it can become a fundamental tool to improve the quality of life of the people around us, not only now but also in the future. Increasingly, as opposed to the traditional success criteria based on standardization, cost efficiency or replicability of solutions, the positive, flexible and creative way in which DT faces challenges and problematic situations of any kind is being positively valued in most working environments. As discussed in this book, the problem-oriented DT approach, an approach that encourages collaboration, concern for human needs, efficiency and sustainability, has brought with it principles and working practices that have proven to be very appropriate in most diverse scenarios.

However, there is no consensus among design thinkers on how DT will develop in the future. At present, many companies and institutions are experimenting with DT, but in many cases, without human resources with the necessary attitudes, experience and skills to contribute to its real success. These entities are faced with the challenges posed by the scope and complexity of today's problems, but it is a real challenge for them to have the time, talent, tools and funding to address them. This situation adds an additional attraction to DT training and in general to training in the soft skills associated with the DT methodology. Having these skills in addition to the specific training as engineers can be an important boost for a professional career, as they will provide access to better job opportunities.

As collaborative work becomes more sophisticated as a consequence of the challenges of today's society, the tools of the DT universe will evolve. Advances in information technology will make digital versions of the most commonly used tools available, facilitating their application at any time by dispersed multidisciplinary teams. Advances in artificial intelligence, human–computer interaction, affective computing or adaptive software, to name a few of the most relevant technologies in this field, will allow a more natural, efficient and effective use, without requiring specific training by design thinkers or end users.

In short, DT tools and techniques will have to be more versatile, human-centered and allow a simple and intuitive application. Currently, the different tools have different origins and original fields of application and are not designed for people coming from different disciplines, so they generally have a not negligible learning curve, especially when they are intended to be used together throughout the DT process. Our methodology will also need standardized and traceable processes to tackle more complex problems, so that a better understanding of the problems and solutions produced can be reached. Integrated DT tool portfolios will be developed, allowing us to move seamlessly from one phase to another and share process results along the way. How DT and its collaborative approach can quickly and effectively solve increasingly complex problem situations will determine the future of the methodology.

References

[1]   Auernhammer J and Roth B. The origin and evolution of Stanford University's design thinking: from product design to design thinking in innovation management. *Journal of Product Innovation Management*. 2021;38(6):623–644.

[2]   Cooley M. Human-centered design. In R. Jacobson (Ed.), *Information Design*. Cambridge, MA: MIT Press, 2000. p. 59–81.

[3]   Rowe PG. *Design Thinking*. Cambridge, MA: MIT Press, 1991.

[4]   Faste R. Ambidextrous thinking. In *Innovations in Mechanical Engineering Curricula for the 1990s*. New York, NY: American Society of Mechanical Engineers, 1994.

[5]   Kelley T and Kelley D. *Creative Confidence: Unleashing the Creative Potential Within Us All*. USA: Currency, 2013.

[6]   Goldman S and Kabayadondo Z. *Taking Design Thinking to the School*. New York, NY: Routledge, Taylor & Francis Group, 2017.

[7]   Levine DI, Agogino AM, and Lesniewski MA. Design thinking in development engineering. *International Journal of Engineering Education*. 2016;32(3):1396–1406.

[8]   Lefebvre RC and Kotler P. Design thinking, demarketing and behavioral economics: fostering interdisciplinary growth in social marketing. In *The Sage Handbook of Social Marketing*. London: Sage, 2011. p. 80–94.

[9]   Knight E, Daymond J, and Paroutis S. Design-led strategy: how to bring design thinking into the art of strategic management. *California Management Review*. 2020;62(2):30–52.

[10]   Koh JHL, Chai CS, Wong B, *et al*. Design thinking and education. In *Design Thinking for Education*. New York, NY: Springer, 2015. p. 1–15.

[11]   Melles G, Howard Z, and Thompson-Whiteside S. Teaching design thinking: expanding horizons in design education. *Procedia-Social and Behavioral Sciences*. 2012;31:162–166.

[12]   Razzouk R and Shute V. What is design thinking and why is it important? *Review of Educational Research*. 2012;82(3):330–348.

[13]   Rogers C and Farson R. *Active Listening*. Grovetown, GA: Mockingbird Press LLC, 2021.

[14]   Vinsel L. Design thinking is kind of like syphilis – it's contagious and rots your brains. *Medium*. December 4, 2017.

[15]   Dewey J. *Art as Experience*. New York, NY: Minton, Balch & Co., 1934.

[16]   von Tury FJ. Design at the crossroads. *American Ceramic Society Bulletin*. 1957;36(12):466–467.

[17] Arnold JE and William JC. *Training Creative Engineers: The Arcturus IV Case Study and its Relation to MDRS.* Boulder, CO: The Mars Society, 2017.

[18] Archer LB. *Systematic Method for Designers.* Council of Industrial Design. UK: H.M.S.O., 1965.

[19] Lawson B. *How Designers Think: The Design Process Demystified.* London: Architectural, 1980.

[20] Cross N, Dorst K, and Roozemburg N. *Research in Design Thinking.* The Netherlands: Delft University Press, 1992.

[21] Buchanan R. Wicked problems in design thinking. *Design Issues.* 1992;8(2): 5–21.

[22] Kelley D and Kelley T. *Creative Confidence: Unleashing the Creative Potential Within Us All.* Glasgow: William Collins Books, 2013.

[23] Clark K and Smith R. Unleashing the power of design thinking. *Design Management Review.* 2008;19(3):8–15.

[24] Parker A, Norman G, Martin B, *et al.* Testing decentralised treatment solutions for portable home toilet waste-Kumasi, Ghana. In Shaw RJ (Ed.), *Proceedings of the 38th WEDC International Conference.* Loughborough, UK: Loughborough University, 2015.

[25] Smith K. From Silicon Valley to Gare du Lyon: SNCF embraces design thinking. *International Railway Journal.* 2017;57(9):50–54.

[26] Murray C. Case study: product innovation at Bank of America. *Commercial Lending Review.* 2009;24:35.

[27] Colin K and Hecht S. *Usefulness in Small Things: Items from the Under a Fiver Collection.* New York, NY: Rizzoli International Publications, 2011.

[28] Thomke S and Feinberg B. *Design Thinking and Innovation at Apple.* Boston, MA: Harvard Business School, 2009. p. 1–12.

[29] Drews C. Unleashing the full potential of design thinking as a business method. *Design Management Review.* 2009;20(3):38–44.

[30] Liedtka J. Innovative ways companies are using design thinking. *Strategy & Leadership.* 2014;42:40–45.

[31] Rogers C. *Way of Being.* Boston, MA: Houghton Mifflin Company, 1995.

[32] Adler A. *What Life Could Mean to You.* Scotts Valley, CA: CreateSpace Independent Publishing Platform, 2015.

[33] Goleman D. *Emotional Intelligence: Why it Can Matter More Than IQ.* London: Bloomsbury Publishing PLC, 1996.

[34] Ohno T and Bodek N. *Toyota Production System: Beyond Large-Scale Production.* New York, NY: Productivity Press, 2019.

[35] Aron A, Melinat E, Aron EN, *et al.* The experimental generation of interpersonal closeness: a procedure and some preliminary findings. *Personality and Social Psychology Bulletin.* 1997;23(4):363–377.

[36] Kolko J. Abductive thinking and sensemaking: the drivers of design Synthesis. *Design Issues.* 2010;26(1):15–28.

[37] Olson T. *Empathetic or Egocentric?* Raleigh, NC: Pendo.io, Inc., 2015. Available from: https://www.pendo.io/pendo-blog/empathetic-or-egocentric/.

[38]   Serrat O. *The Five Whys Technique*. Gladstone, NJ: Knowledge Solutions. 2009. p. 30.

[39]   Rawlinson JG. *Creative Thinking and Brainstorming*. London and New York, NY: Routledge (Taylor & Francis Group), 2017.

[40]   Mateus R, Neiva S, Bragança L, *et al*. Sustainability assessment of an innovative lightweight building technology for partition walls – comparison with conventional technologies. *Building and Environment*. 2013;67: 147–159. Available from: http://www.sciencedirect.com/science/article/pii/ S0360132313001571.

[41]   Daidj N. Uberization (or Uberification) of the Economy. In Khosrow-Pour M (Ed.), *Advanced Methodologies and Technologies in Digital Marketing and Entrepreneurship*. Hershey, PA: IGI Global, 2019. p. 116–128.

[42]   Bjelicic B. The business model of low cost airlines. Past, present, future. In Groß S and Schröder A (Eds), *Handbook of Low Cost Airlines: Strategies, Business Processes and Market Environment*. Berlin: Eric Schmidt Verlag, 2007. p. 11–30.

[43]   Nenonen S and Storbacka K. *SMASH: Using Market Shaping to Design New Strategies for Innovation, Value Creation, and Growth*. Bingley: Emerald Publishing, 2018.

[44]   Kristensson P, Magnusson P, and Witell L (Eds). *Service Innovation for Sustainable Business*. Singapore: World Scientific, 2019.

[45]   Gryskiewicz S and Taylor S. *Making Creativity Practical: Innovation That Gets Results*. Greensboro, NC: Center for Creative Leadership, 2003.

[46]   Buzan T. *Use Your Head*. London: Gild Publishing, 1974.

[47]   Hillar SP. *Mind Mapping with FreeMind*. Birmingham: Packt Publishing, 2012.

[48]   Cameron H and Voight R. *MindManager for Dummies*. New York, NY: Wiley Publishing, 2004.

[49]   Caeiro Rodríguez M and Fernández Iglesias MJ. *Buscando soluciones innovadoras: de la tormenta de ideas al desarrollo de concepto*. Departamento de Ingeníeria Telemática. Universidade de Vigo, 2020. M3HG6. Available from: https://osf.io/M3HG6.

[50]   Beaudouin-Lafon M and Mackay W. Prototyping tools and techniques. In Jacko JA and Andrew Sears A (Eds), *The Human–Computer Interaction Handbook*. Mahwah, NJ: L. Erlbaum Associates Inc., 2002. p. 1006–1031.

[51]   Still B and Morris J. The blank-page technique: reinvigorating paper prototyping in usability testing. *IEEE Transactions on Professional Communication*. 2010;53(2):144–157.

[52]   Halskov K and Nielsen R. Virtual Video Prototyping. *Human–Computer Interaction*. 2006;21(2):199–233. Available from: https://www.tandfonline.com/ doi/abs/10.1207/s15327051hci2102_2.

[53]   Robbins L. A whole new game ball? N.B.A. admits its mistake. *The New York Times*. December 6, 2006. Available from: https://www.nytimes.com/2006/ 12/06/sports/basketball/06ball.html.

[54]   Vertelney L. Using video to prototype user interfaces. *SIGCHI Bulletin* 1989;21(2):57–61. Available from: http://doi.acm.org/10.1145/70609.70615.

[55] Owens J. *Video Production Handbook*, 6th ed. New York, NY: Routledge and Taylor & Francis Group, 2017.

[56] Schlögl S, Doherty G, Karamanis N, *et al*. WebWOZ: a wizard of Oz prototyping framework. In *Proceedings of the 2nd ACM SIGCHI Symposium on Engineering Interactive Computing Systems. EICS '10*. New York, NY: Association for Computing Machinery, 2010. p. 109–114. Available from: https://doi.org/10.1145/1822018.1822035.

[57] Maulsby D, Greenberg S, and Mander R. Prototyping an intelligent agent through wizard of Oz. In *Proceedings of the INTERACT '93 and CHI '93 Conference on Human Factors in Computing Systems. CHI '93*. New York, NY: Association for Computing Machinery, 1993. p. 277–284. Available from: https://doi.org/10.1145/169059.169215.

[58] Beyer AM. Improving student presentations: Pecha Kucha and just plain PowerPoint. *Teaching of Psychology*. 2011;38(2):122–126.

[59] Carroll M, Goldman S, Britos L, *et al*. Destination, imagination and the fires with design thinking in a middle school classroom. *International Journal of Art & Design Education*. 2010;29(1):37–53.

[60] Monserrat JF, Diehl A, Bellas Lamas C, *et al. Envisioning 5G-Enabled Transport*. Washington, DC: World Bank, 2020.

[61] Novickis R, Levinskis A, Kadikis R, *et al*. Functional architecture for autonomous driving and its implementation. In *Proceedings of the Biennial Baltic Electronics Conference, BEC*, October 2020.

[62] Simoudis E. The autonomous mobility innovation lifecycle. *IEEE Potentials*. 2020;39(1):9–14.

[63] Frey T. *Driverless Tech – 8 Scenarios That Show it to be the Most Disruptive Technology in All History*, 2017. Available from: https://futuristspeaker.com/.

[64] Townsend RM, Atkinson-Palombo C, Terbeck F, *et al*. Hopes and fears about autonomous vehicles. *Case Studies on Transport Policy*. 2021;9(4): 1933–1942. Available from: https://www.sciencedirect.com/science/article/pii/S2213624X21001760.

[65] Townsend A. *Re-programming Mobility: The Digital Transformation of Transportation in the United States*. New York, NY: Rudin Center for Transportation Policy and Management, 2014.

[66] Cassetta E, Marra A, Pozzi C, *et al*. Emerging technological trajectories and new mobility solutions. A large-scale investigation on transport-related innovative start-ups and implications for policy. *Transportation Research Part A: Policy and Practice*. 2017;106:1–11. Available from: https://www.sciencedirect.com/science/article/pii/S0965856417302987.

[67] Pigeon C, Alauzet A, and Paire-Ficout L. Factors of acceptability, acceptance and usage for non-rail autonomous public transport vehicles: a systematic literature review. *Transportation Research Part F: Traffic Psychology and Behaviour*. 2021;81:251–270. Available from: https://www.sciencedirect.com/science/article/pii/S1369847821001480.

[68] Tan H, Zhao X, and Yang J. Exploring the influence of anxiety, pleasure and subjective knowledge on public acceptance of fully autonomous

vehicles. *Computers in Human Behavior*. 2022;131:107187. Available from: https://www.sciencedirect.com/science/article/pii/S0747563222000097.

[69]   Longo M, Yaïci W, and Foiadelli F. Future mobility advances and trends. In Găiceanu M (Ed.), *Self-Driving Vehicles and Enabling Technologies*. Rijeka: IntechOpen, 2021. Available from: https://doi.org/10.5772/intechopen.97108.

[70]   Kaur K. A survey on internet of things – architecture, applications, and future trends. In *2018 First International Conference on Secure Cyber Computing and Communication (ICSCCC)*, 2018. p. 581–583.

[71]   Brincat AA, Pacifici F, Martinaglia S, *et al.* The Internet of Things for intelligent transportation systems in real smart cities scenarios. In *IEEE 5th World Forum on Internet of Things, WF-IoT 2019 – Conference Proceedings*, 2019. p. 128–132.

[72]   SAE International. *Taxonomy and Definitions for Terms Related to Driving Automation Systems for On-Road Motor Vehicles*, 2021. Available from: https://www.sae.org/standards.

[73]   Wellbrock W, Ludin D, Röhrle L, *et al.* Sustainability in the automotive industry, importance of and impact on automobile interior – insights from an empirical survey. *International Journal of Corporate Social Responsibility*. 2020;5(1): 1–11.

[74]   Munoz F. The Global Electric Car Sales 2021 in Numbers, 2022. Available from: https://www.jato.com/the-global-electric-car-sales-2021-in-numbers/.

[75]   blog J. *EVs Outsell Diesel Vehicles in Europe in August for the First Time Ever*. Available from: https://www.jato.com/evs-outsell-diesel-vehicles-in-europe-in-august-for-the-first-time-ever/.

[76]   Altshuller G. *40 Principles: TRIZ Keys to Innovation*, vol. 1. Worcester, MA: Technical Innovation Center, Inc., 2002.

[77]   Frey CB and Osborne M. The future of employment: how susceptible are jobs to computerisation? *Technological Forecasting & Social Change*. 2013;114: 254–280. Oxford Martin Programme on Technology and Employment.

Index